Per Ardua Libertas

S.O.A. BOOKS

Published by S.O.A. Books, Canberra

First promulgated by M.I.9 in 1942

ISBN: **978-1-925907-09-4**

This edition © S.O.A. Books, 2024, all rights reserved. No part of this publication may be reproduced, stored in a retrieval system or transmitted, in any form or by any means, electronic, mechanical, photocopying, recording or otherwise without prior permission from the publisher.

MOST SECRET

THE FUNCTIONS OF M.I.9 WERE GOVERNED BY A CHARTER TO SERVE ══

—— THE NAVY ——
—— THE ARMY ——
THE ROYAL AIR FORCE

M. I. 9
TECHNICAL

**WAR OFFICE
LONDON
S.W.1**

War Office,

London, S.W.1.

14.2.1942.

NOTE.

The following pages give a photographic review of the range of work I was privileged to be entrusted with on behalf of Section M.I.9, for two years between February 14th, 1940, and February 14th, 1942.

They show "Aids to Escape" - Pre-Capture and Post-Capture - and in addition show various other articles called for by various Sections of the three Services which, through the channels laid on, M.I.9, were enabled to produce or deliver quickly.

No details are given in this review of the difficulties experienced in obtaining manufacture of the various Aids. These are dealt with in separate notes—as are the names of the manufacturers.

Two points, perhaps, should be put on record. During the period covered, no finished working suggestion was ever submitted to me by any other Service Department and no Service Factory or Organisation was used in the manufacture of any article.

I should like to record my sincere thanks to Colonel N. R. Crockatt, D.S.O., M.C., for his kindly understanding of the very difficult problems with which I was faced and for the considerable latitude he has always granted me in letting me work in my own irregular way; without such help the results shown in this book could never have been so effectively achieved.

With but few exceptions, all articles were devised and production obtained by me.

Major.

The Charter.

CONDUCT OF WORK No. 48.
M.I.9.

1. A new section of the Intelligence Directorate at the War Office has been formed. It will be called M.I.9. It will work in close connection with and act as agent for the Admiralty and Air Ministry.

2. The Section is responsible for:—

   (a) The preparation and *execution* of plans for facilitating the escape of British Prisoners of War of all three Services in Germany or elsewhere.

   (b) Arranging instruction in connection with above.

   (c) Making other advance provision, as considered necessary.

   (d) Collection and dissemination of information obtained from British Prisoners of War.

   (e) Advising on counter-escape measures for German Prisoners of War in Great Britain, if requested to do so.

3. M.I.9. will be accommodated in Room 424, Metropole Hotel.

(Sgd.) J. SPENCER.
Col. G.S.
for D.M.I.

23.12.39.

On February 14th 1941 the Commissioned Staff
of M.I.9 consisted of:

COLONEL N. R. CROCKATT, D.S.O., M.C.

MAJOR V. A. R. ISHAM, M.C.   MAJOR C. M. RAIT, M.C.

SQDN. LEADER A. J. EVANS, M.C.   COMMANDER P. W. RHODES, R.N.

MAJOR C. CLAYTON HUTTON
(TECHNICAL)

●

OPERATIONS WERE CONDUCTED FROM THE
WAR OFFICE   LONDON, S.W.1.

●

MAPS

After exhaustive tests to find a material that would be waterproof, subject to the smallest printing and confinable in the smallest space, a special type of silk was used.

At first this material was printed on one side only, but it was found that by special treatment maps could be printed on both sides of it.

Later on, artificial silk was similarly treated and used as well.

Up to February 14th, 1942, 56 maps were produced covering every theatre of war as the following lists show.

Portions of some printed specimens on double-sided and one-sided silk are shown on pages 20—23.

THE FOLLOWING MAPS WERE PRINTED ON ONE-SIDED SILK AND ON PAPER.

Germany (**5** *kinds*).
North France (**2** *kinds*).
South France (**2** *kinds*).
England and **North** France.
Norway and Sweden (**2** *kinds*).
Norway.
Sweden.
Spain.
Spain, **P**ortugal and **Cor**sica.
North Italy (**2** *kinds*).
South Italy (**2** *kinds*).
Cyrenaica.
Eritrea.
Abyssinia.
Somaliland.
Juba River.

Roumania and Bulgaria (**2** *kinds*).
Russia.
North West Russia, Poland, Finland.
Greece and Jugoslavia (**2** *kinds*).
Turkey.
Caucasus.
Syria and Iraq.
Persia.
Middle East.
Iran.
Spittal (*route*).
Basle (*route*).
South Germany (*route*).
Schaffhausen (*route*) (**2** *kinds*).
Baltic (*route*).
Brussels (3 *parts*).

MAPS

THE FOLLOWING MAPS WERE PRINTED ON DOUBLE-SIDED SILK AND ON PAPER.

Germany   North France.

Germany   **North** France and England.

Germany   Norway and Sweden.

North France   **South** France.

South France   Spain.

South France   Spain, **Portugal**, and Corsica.

Norway   Sweden.

Spain, Portugal, **Corsica**   North West African Coast.

North Italy   **South** Italy.

South Italy   Cyrenaica.

Cyrenaica   Middle East.

Morocco, Tripoli   North West African **Coast.**

Cyrenaica   Morocco, Tripoli.

North **West** African Coast, **both** sides.

West African Coast, both sides.

Eritrea   Abyssinia.

Roumania, Bulgaria   Greece and Jugoslavia.

Roumania, **Bulgaria**   Russia

Roumania, Bulgaria   Middle East.

Russia   North West Russia, **Pol**and, Finland.

Greece and Crete   Italy, Greece, Turkey in Europe.

Turkey   **Syr**ia, Iraq.

Caucasus   Persia.

Middle East   Iran.

Spittal   Basle.

209,000
MAPS

AND

214,000
AIDS ITEMS

FOR

PRE AND POST CAPTURE

WERE DISTRIBUTED
UP TO FEBRUARY 14th, 1942
TO UNITS OF THE THREE SERVICES

MAKING A TOTAL OF

423,000
AIDS

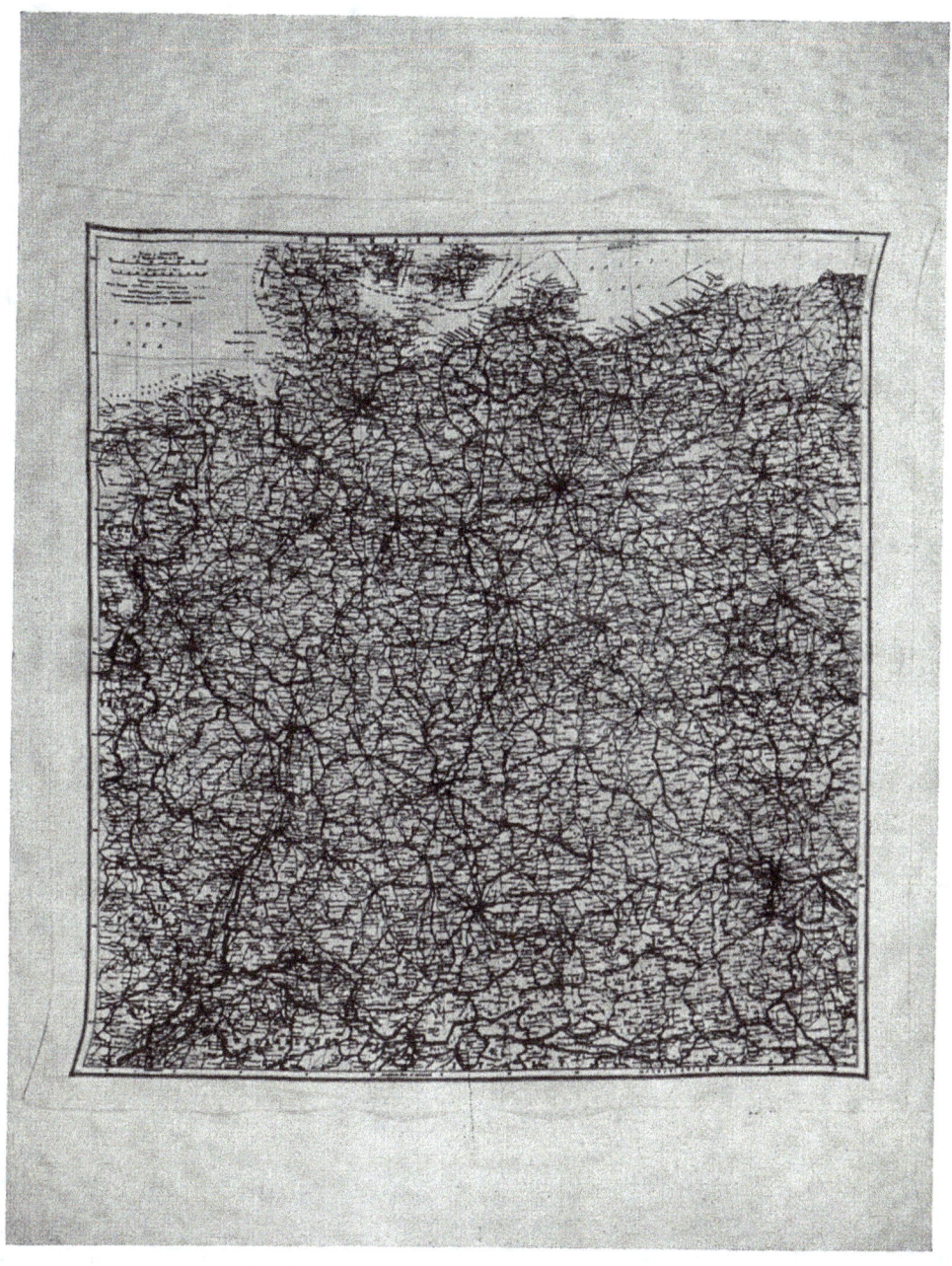

Map of Germany printed in three colours on silk.

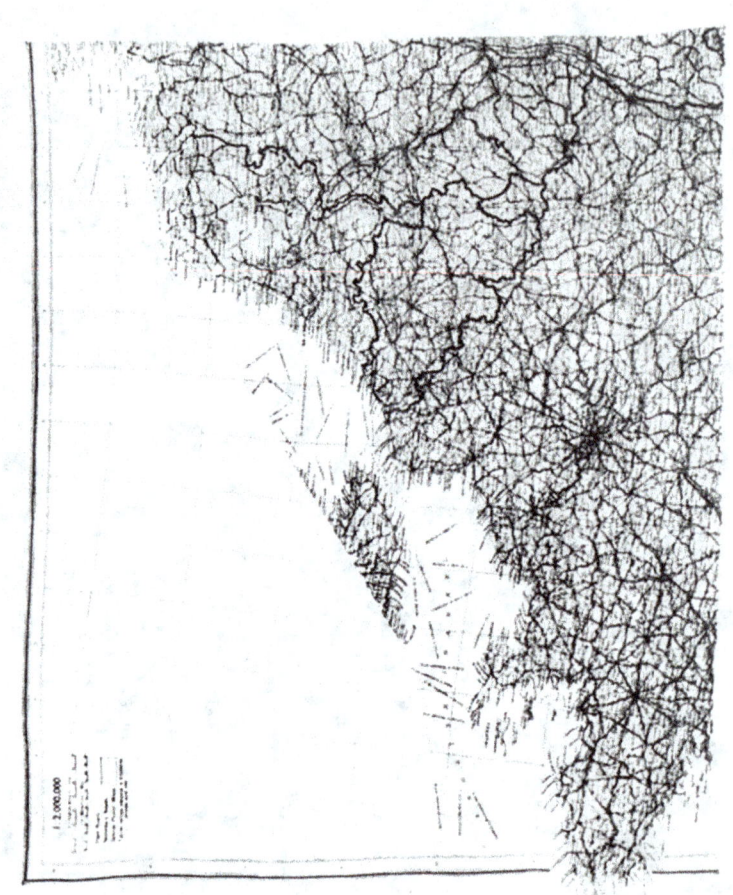

Map of N. France printed in three colours on silk.

Map of Germany in three colours on one side of the silk and
Map of certain German border districts in three colours on reverse.

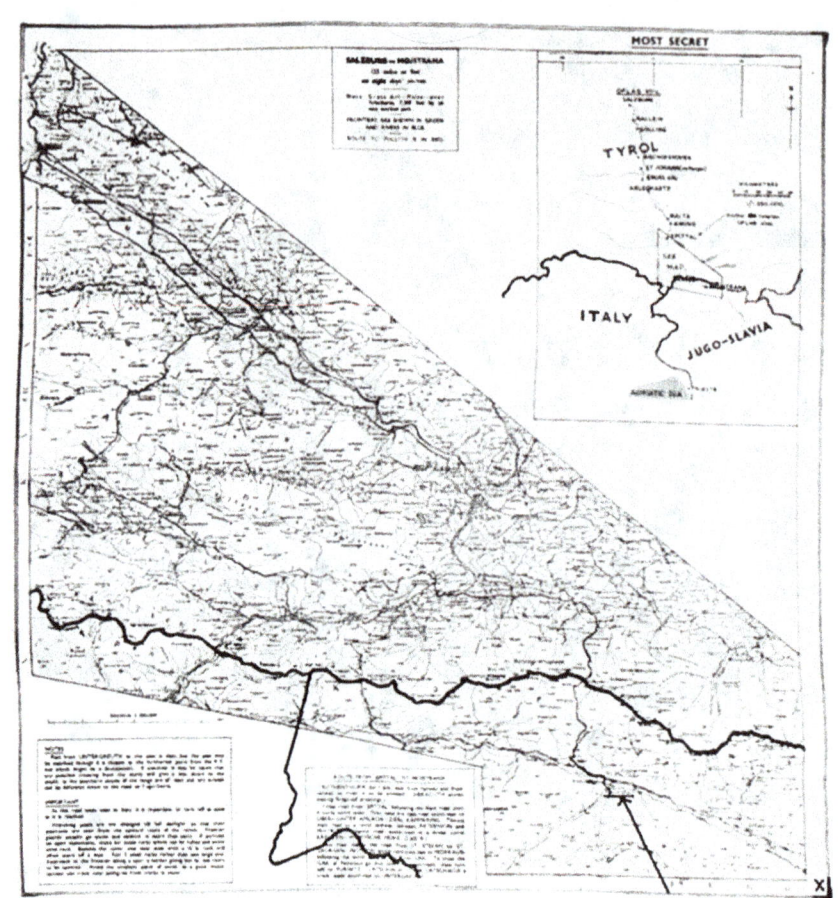

Spittal Route Map printed in four colours on tissue.

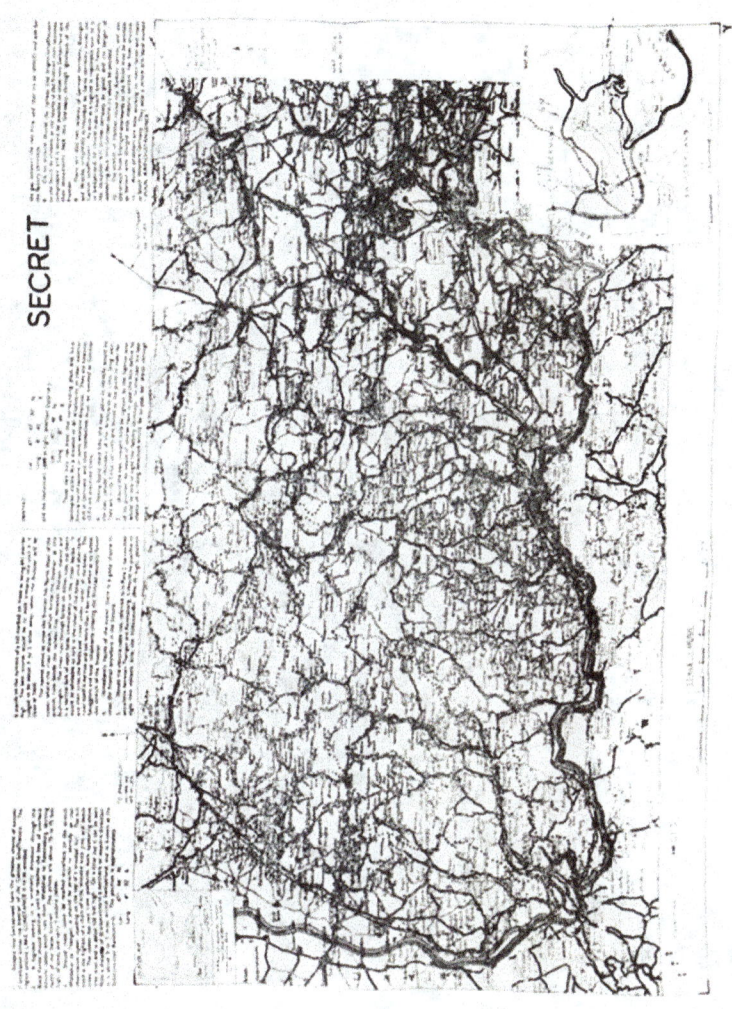

Schaffhausen Route Map printed in four colours on tissue.

SILK MAPS

Single-sided

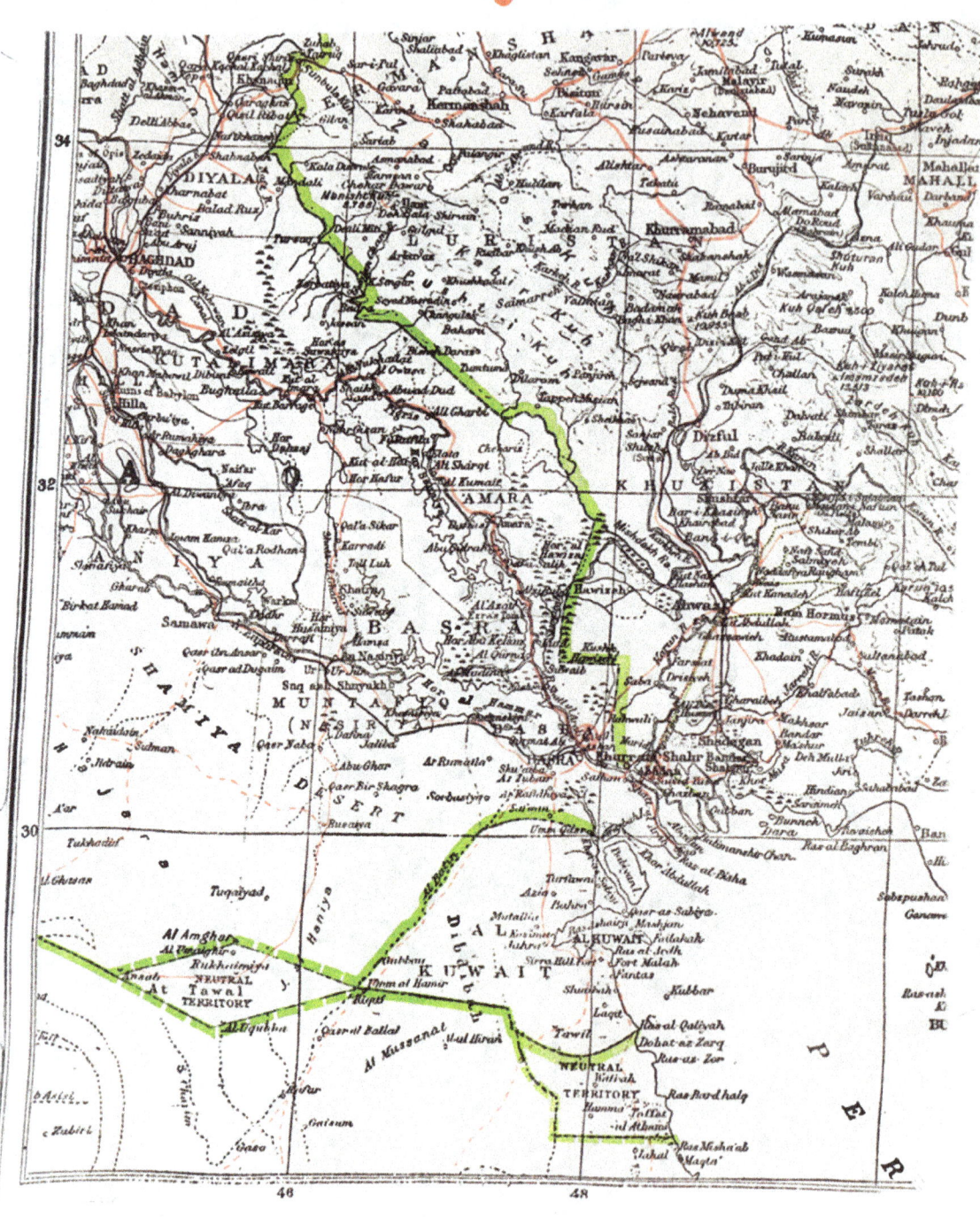

SILK MAPS

Double-sided

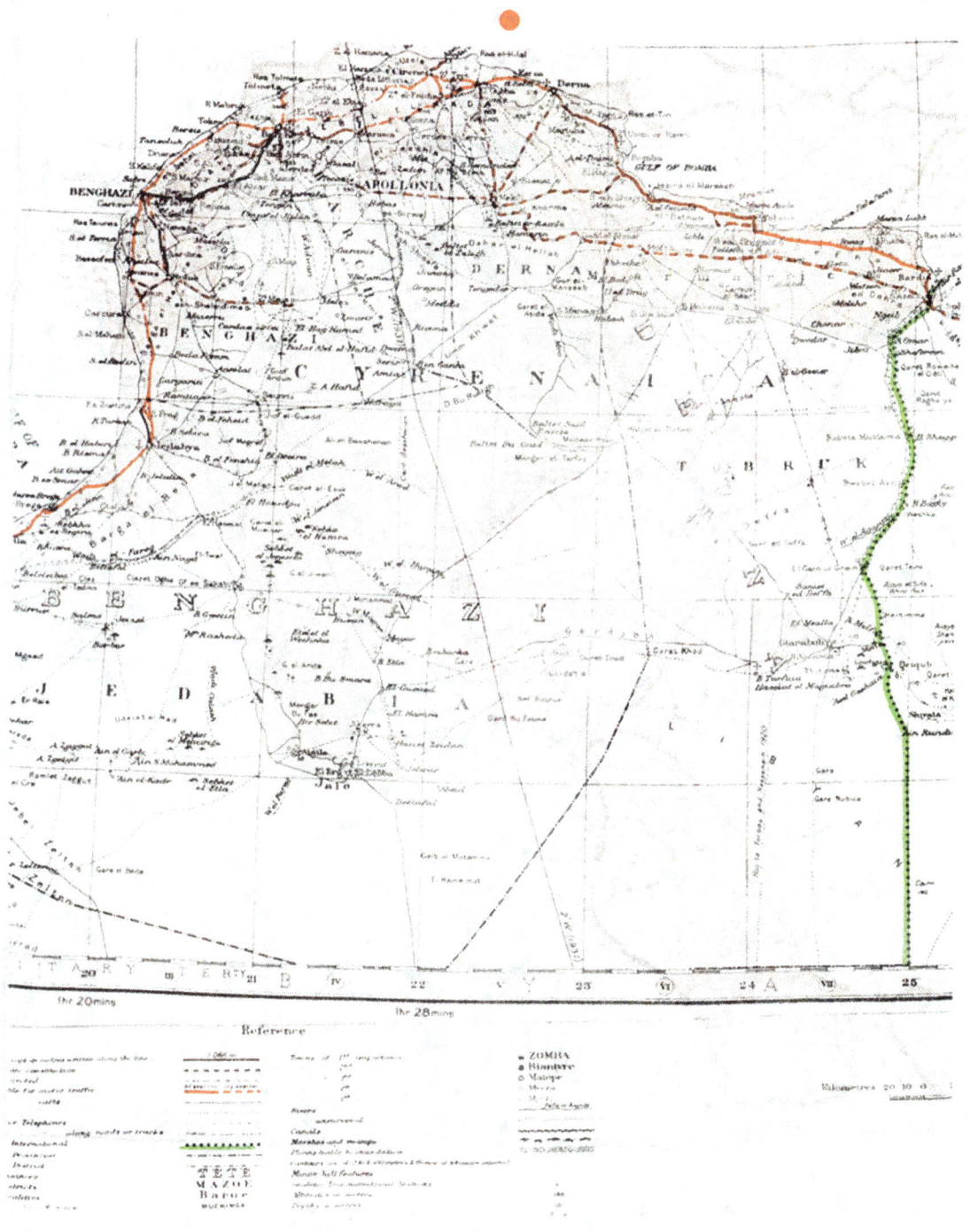

ARTIFICIAL SILK MAPS

(Waterproofed 'Tenasco')

Double-sided

TISSUE MAPS

One-sided only

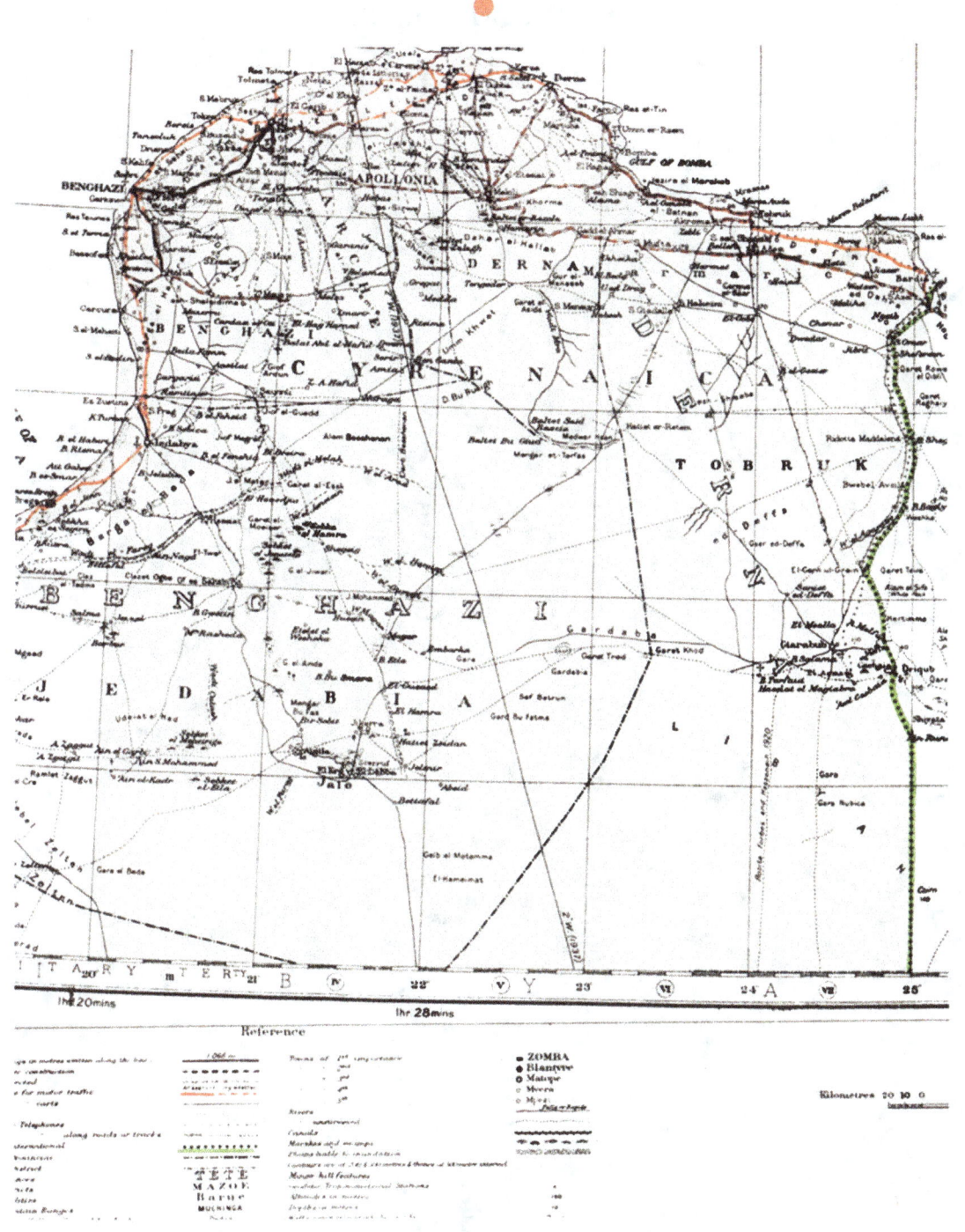

ESCAPE AIDS

A STANDARD ROTOR WAS EVOLVED WHICH COULD BE INTERCHANGED IN VARIOUS ARTICLES.

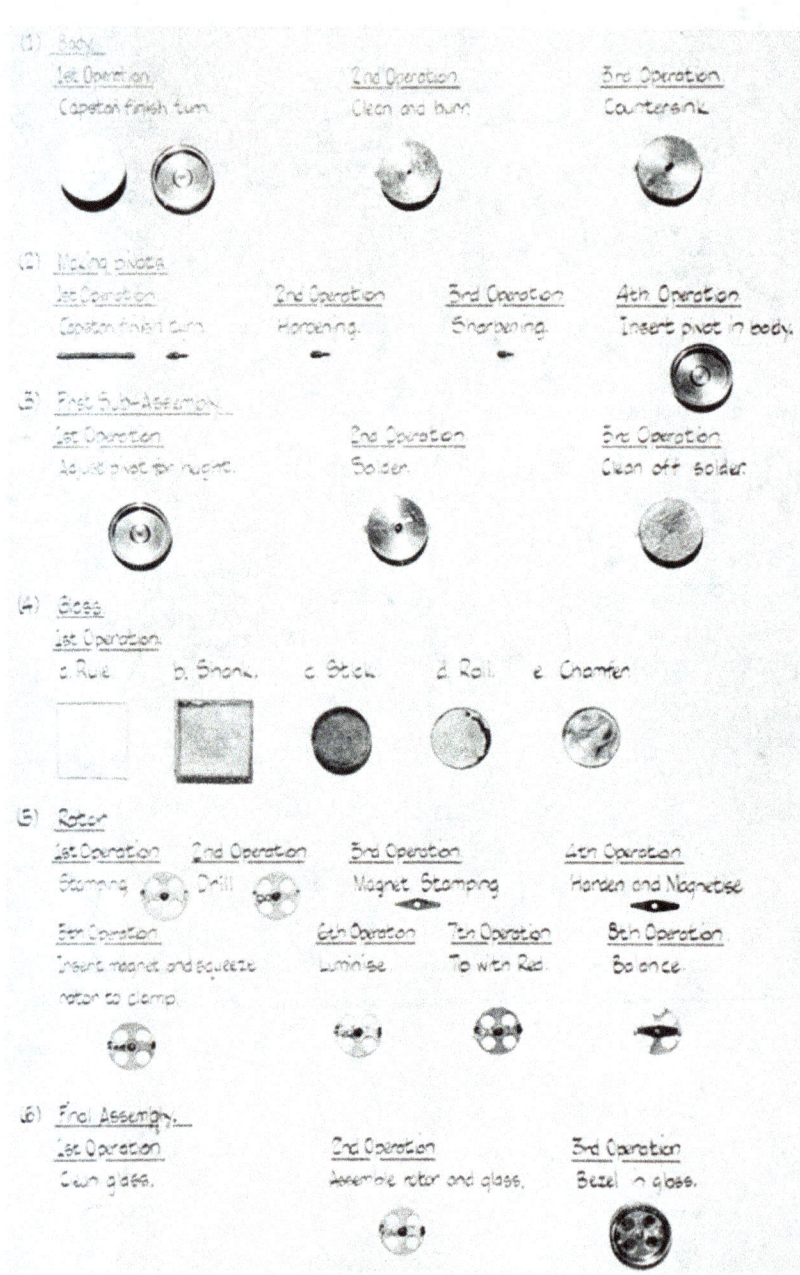

This is photographic lay-out showing complete operation from the raw material to the finished round compass.

ESCAPE AIDS

STANDARD COMPASSES
photographed half size larger than actual.

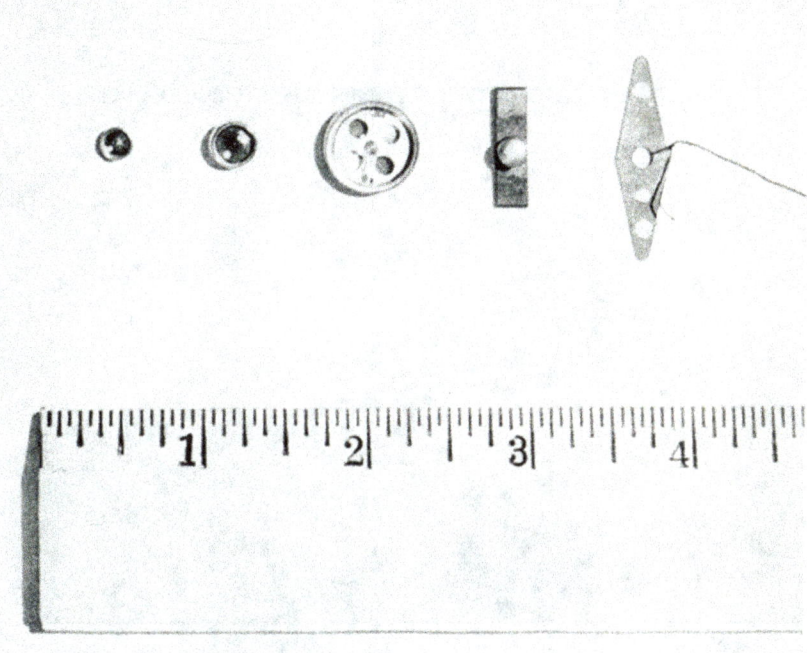

Several types of Compasses were designed, but those most used consisted of various sizes of

Luminous Rotating
Hanging Swinging
Balancing Swinging

PRE-CAPTURE

STANDARD R.A.F. BUTTON LUMINOUS COMPASS.

Same design applied to all Service Buttons.

STANDARD WOOLWORTH LUMINOUS STUD COMPASS

ORDINARY BRASS SERVICE FLY BUTTON.   (In pairs.)

One Button swings on top of the other, pointing due north.
Luminous.

FOUNTAIN PEN COMPASS

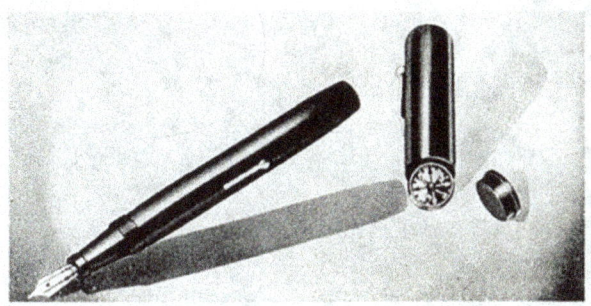

Various types of this were used.

PIPE

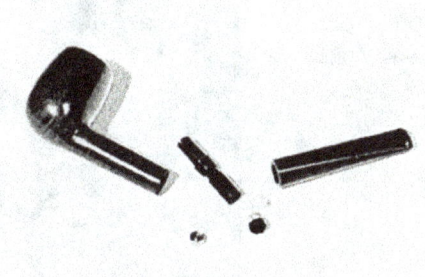

This contained compass and also a chemical substance.

ESCAPE AIDS

RATION BOX. 1.

(Size 5¾" x 4¼" x ¾")

This Ration Box is in a 50 'Players' Cigarette Tin.

Of this model and model 2 there were supplied up to February 14th, 1942 — 23,800

Contents.

1. Rubber Water Bottle.
2. Horlicks Tablets.
3. Packet of Benzedrine Drugs.
4. Packet of Halazone Water Softener.
5. Packet of Chewing Gum.
6. Bar of Chocolate.
7. Box of Matches.

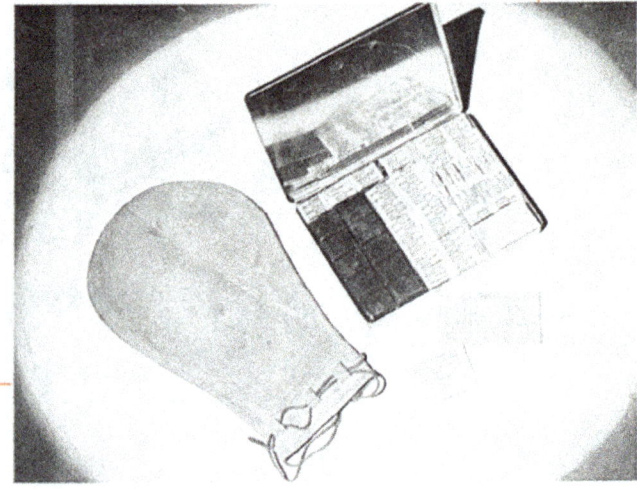

ESCAPE AIDS

RATION BOX. 2.

(Size 6" x 4¼" x 1")

This Ration Box is made of Acetate.

It is transparent and watertight.

Contents.

1. Rubber Water Bottle.
2. Horlicks Tablets.
3. Packet of Benzedrine Drugs.
4. Packet of Halazone Water Softener.
5. Packet of Chewing Gum.
6. Bar of Chocolate.
7. Box of Matches.
8. Two Tubes Condensed Milk.
9. Saw.
10. Compass.

(In *some instances* special Maps were *also* included).

PRE-CAPTURE

HACK-SAW

Hack-saw supplied with straight handle.

Handle bends at given points.

Saw springs into frame.

Handle magnetized to swing due North.

PRE-CAPTURE & POST CAPTURE

RAZOR BLADES

All types were used. Only one blade in each packet magnetized.

Clipping from Blade centre also magnetized.

PRE-CAPTURE & POST CAPTURE

OFFICER'S HAT SUPPORT.

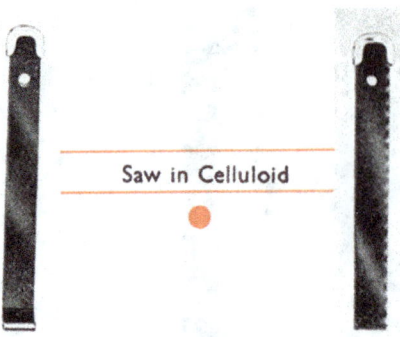

Saw in Celluloid

Fastest and steadiest way of resting Blade and Swinger Compasses — in water.

PRE-CAPTURE & POST CAPTURE

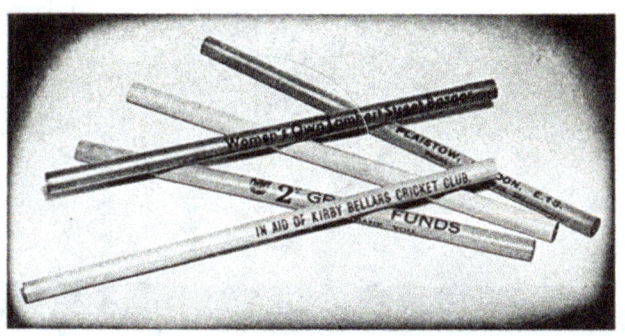

STANDARD SIZE PENCILS.

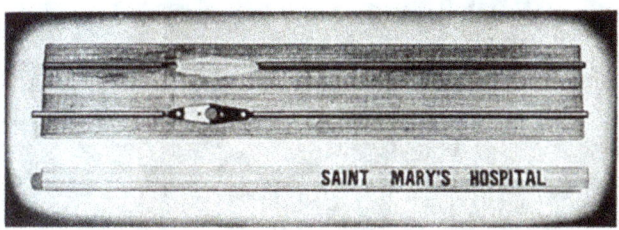

PENCIL CONCEALING SWINGER.

PENCIL CONCEALING THIN SAW IN LEAD ITSELF.

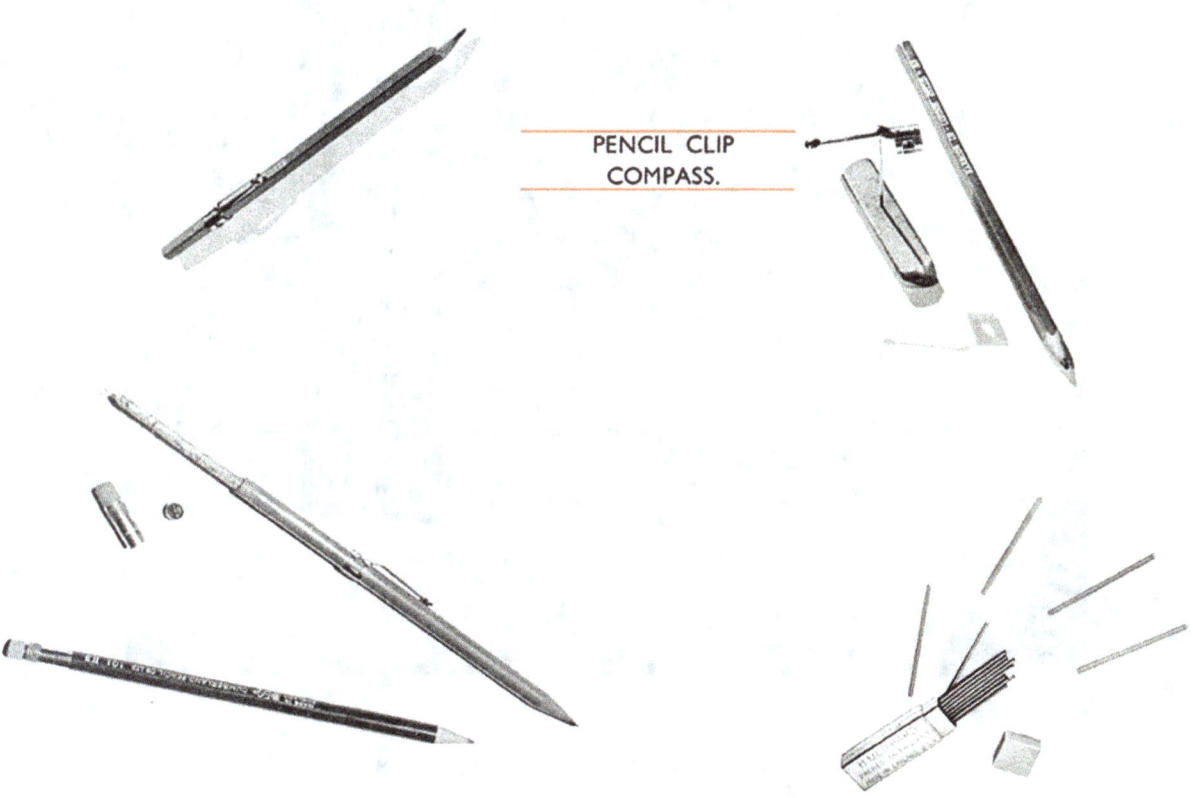

PENCIL CLIP COMPASS.

Contents
1 Map   1 Compass (Luminous)   1 Pencil Clip Compass

STANDARD EVERSHARP PENCIL REFILLS.
One steel lead Compass Needle in each box.

POST CAPTURE

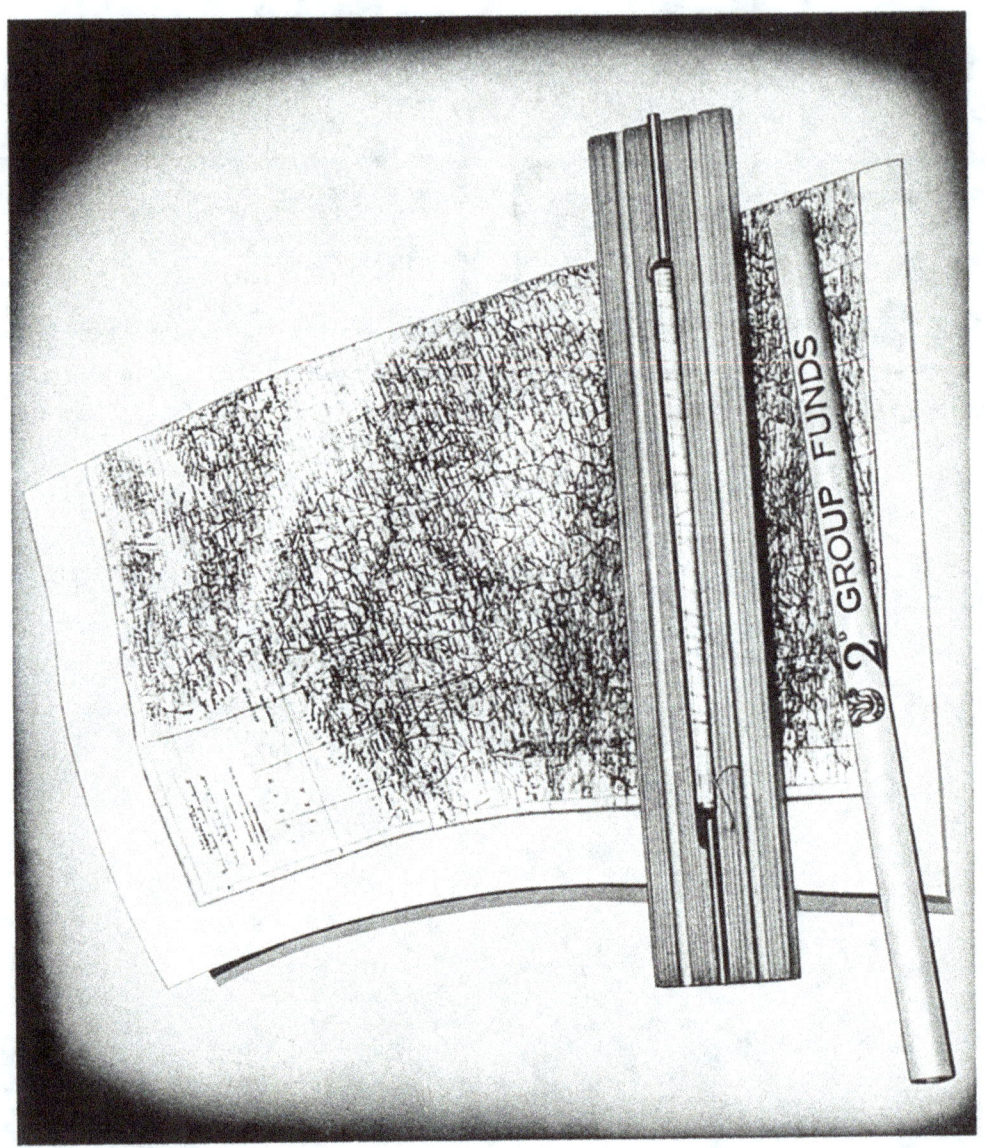

PENCIL CONCEALING ROLLED TISSUE MAP

PRE-CAPTURE & POST CAPTURE

SPECIAL R.A.F. BOOT.

- Boot designed to fit snugly round leg and foot. In boot strap is Knife. By cutting round seam, top part comes away — leaving perfect walking shoe.

POST CAPTURE

GAMES CARRIERS AND THEIR CONTENTS.
(Maps and various escape aids)

SMALL CHESS SET.

CRIBBAGE BOARD.

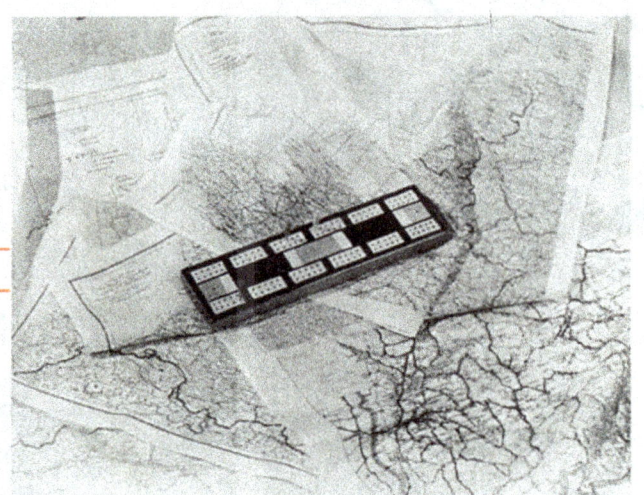

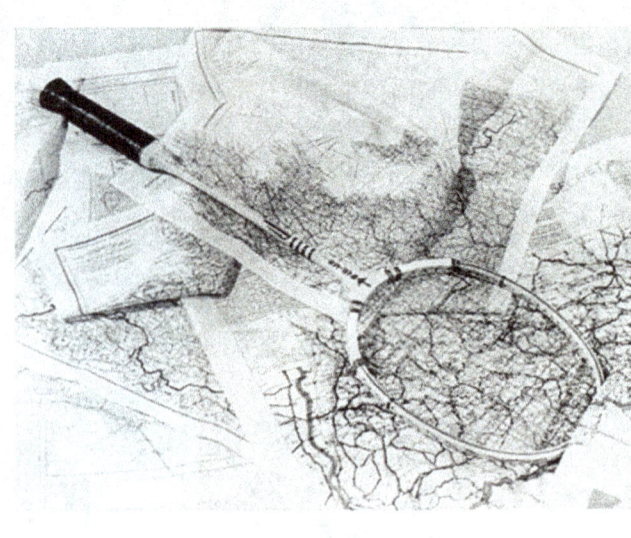

SQUASH RACKET.

POST CAPTURE

GAMES CARRIERS AND THEIR CONTENTS.
(Maps and various escape aids)

MEDIUM CHESS SET.

BACKGAMMON SET.

LARGE CHESS SET.

37

POST CAPTURE

GAMES CARRIERS AND THEIR CONTENTS.
(Maps and various escape aids).

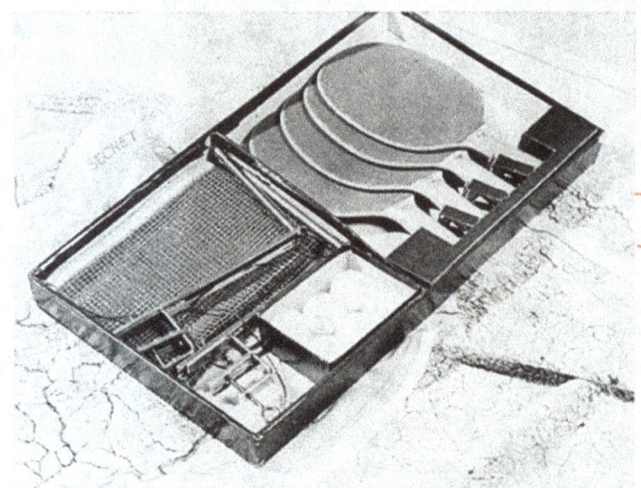

TABLE TENNIS SET.

DRAUGHTS BOARD.

DOMINOES SET.

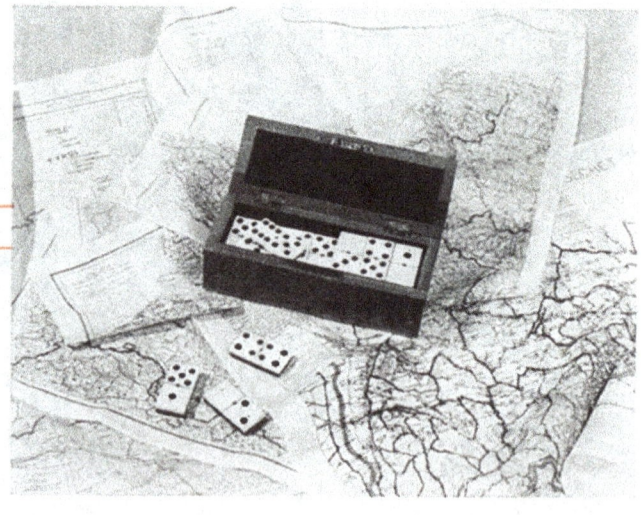

GAMES CARRIERS.

DARTBOARD.

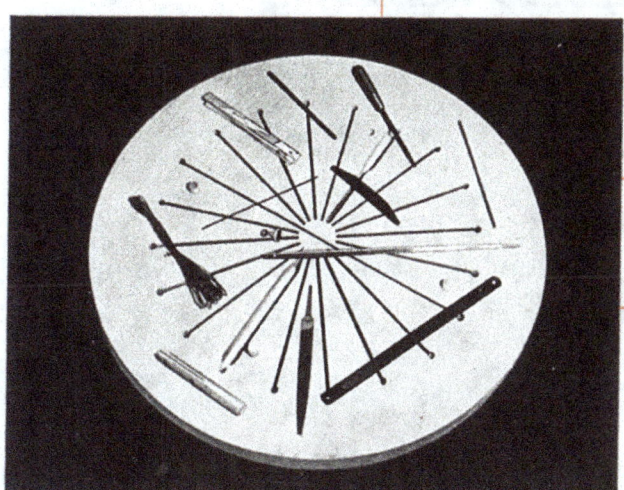

Each game was made by hand by the finest craftsmen, containing various escape aids.

"Doves" were used as well.

POST CAPTURE

WOODEN CARRIERS.

Various wooden carriers used.

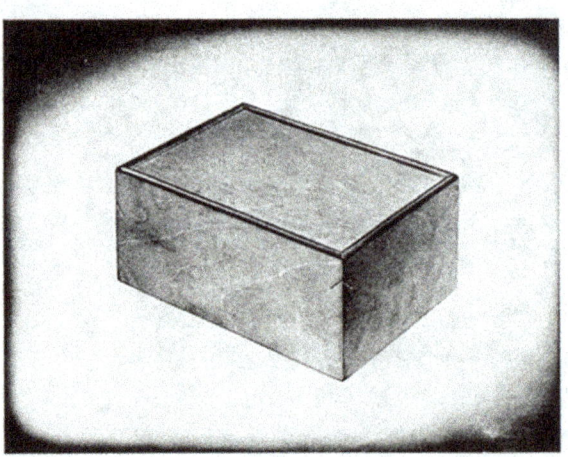

All were made by hand by the finest craftsmen, all elderly men. They were all non subject to X-Ray.

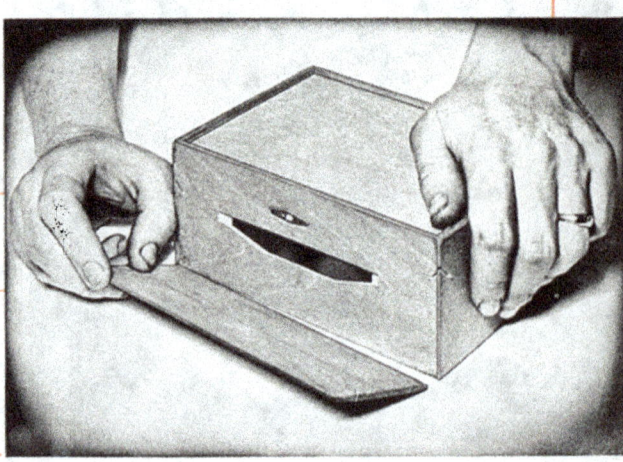

POST CAPTURE

TOILET SET CARRIERS AND THEIR CONTENTS.

(Maps and various escape aids).

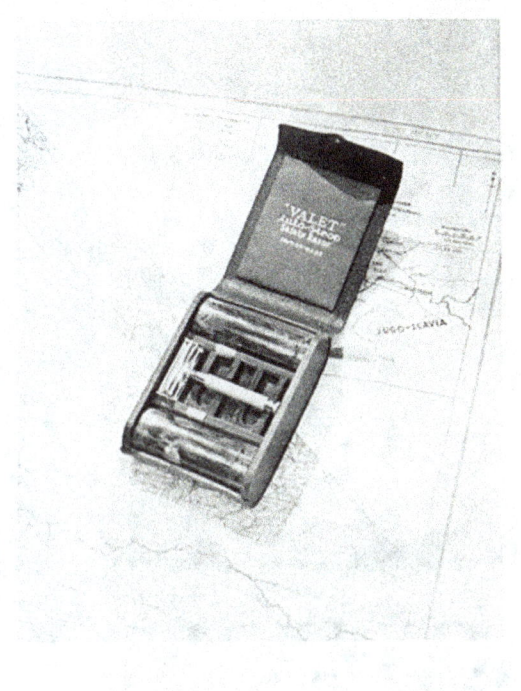

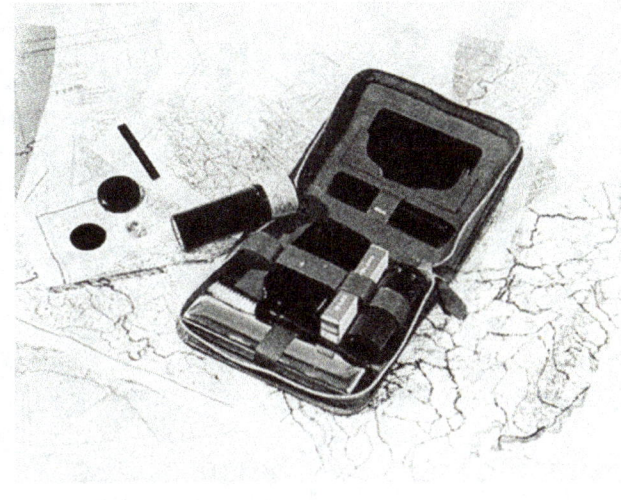

POST CAPTURE

MISCELLANEOUS CARRIERS.

STANDARD PACK OF PLAYING CARDS.
Each pack is one Map. 48 Cards covered a Map. The 4 Aces are a small Map of Europe. The Joker is the Key. The outside Card contains the instructions.

CIGAR CARRIER.
Contains either tissue or silk map and compass.

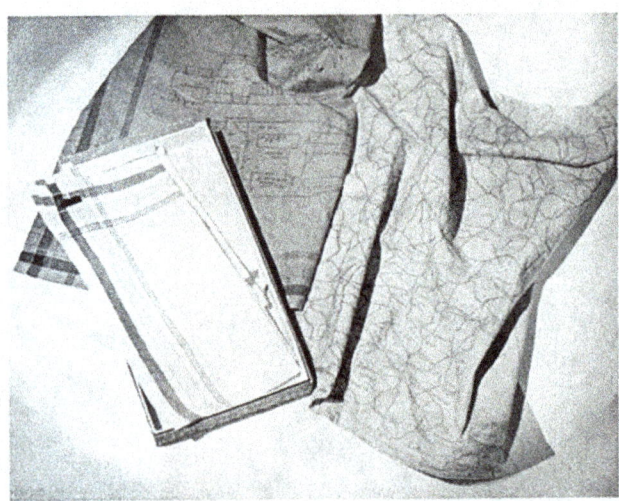

STANDARD COTTON HANDKERCHIEFS.
(Result obtained when washed with chemical on page 65.)

NEW AND SECOND HAND BOOK CARRIER.

POST CAPTURE

MISCELLANEOUS CARRIERS

TOOTH — GOLD FITTING made to measure

Small medium luminous compass fits in jaws on left and thin gold tube holding message or map slides on to the two prongs at bottom. These are concealed through being in between the cheek and gum.

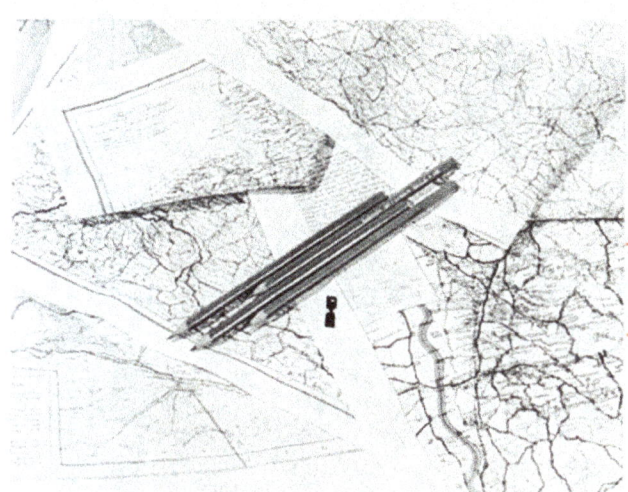

BRIDGE MARKER PENCILS AND THEIR CONTENTS.

STANDARD SANDALS

which were asked for under the name of "Picer" model.

(*Pice being Indian currency.*)

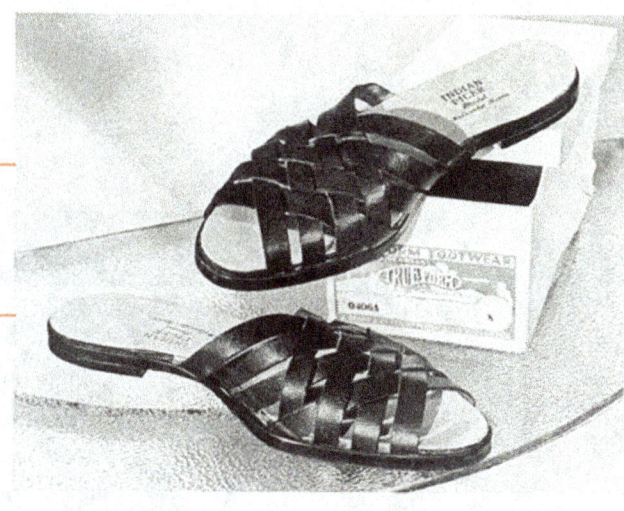

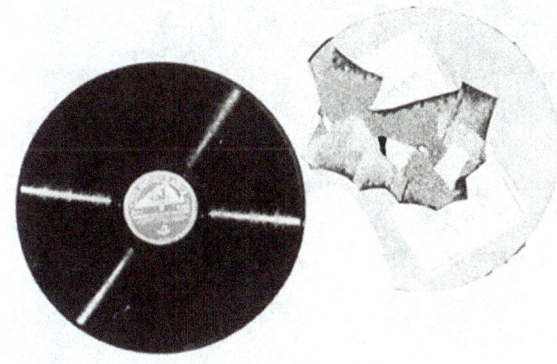

GRAMOPHONE RECORD CARRIER.

This is a standard record and contents were secreted in *both* sides. The record was perfect in every way and could be played.

POST CAPTURE

STANDARD "HALEX" TOILET GOODS.

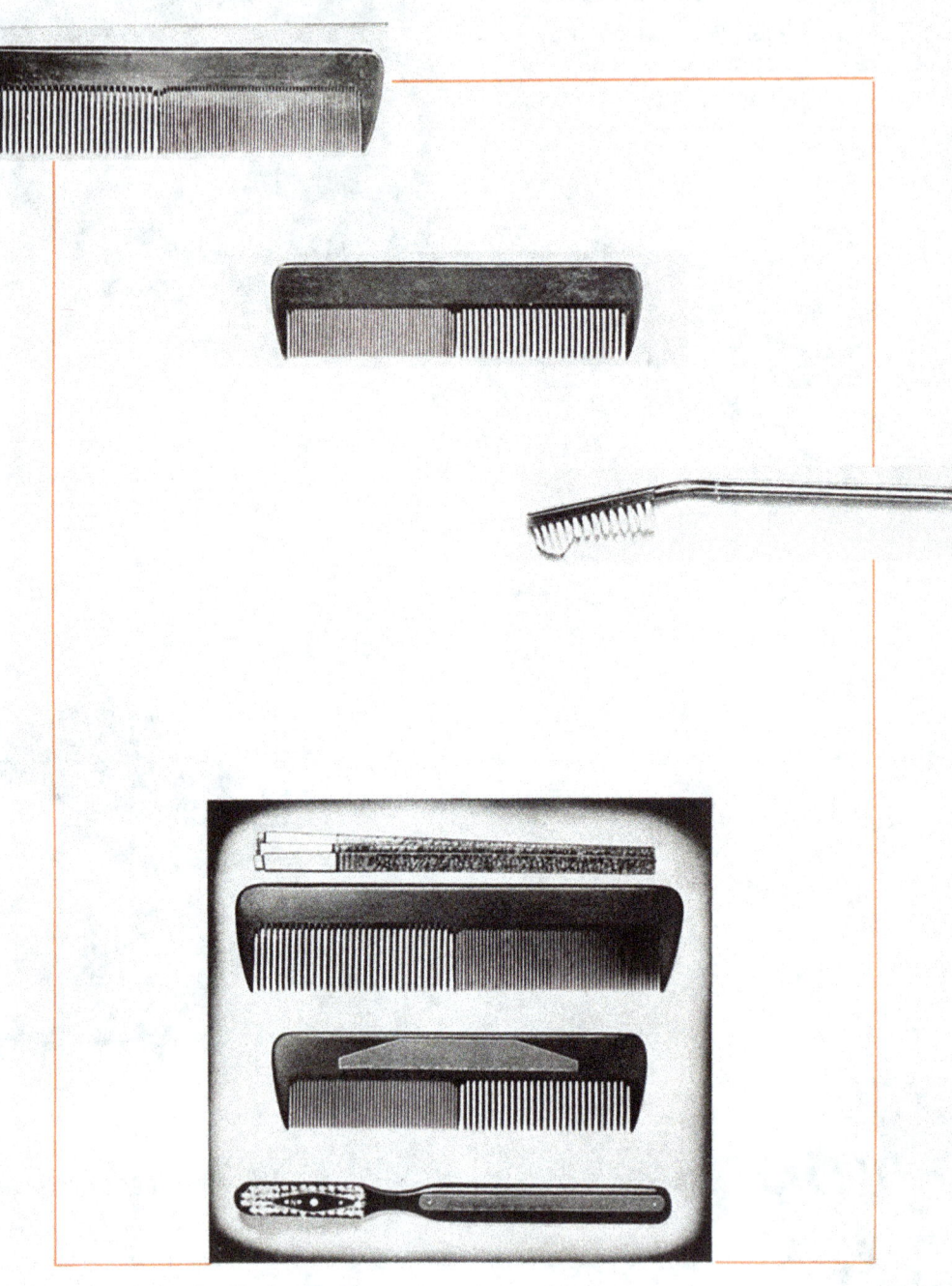

Containing Maps, Saws and Compasses.

POST CAPTURE

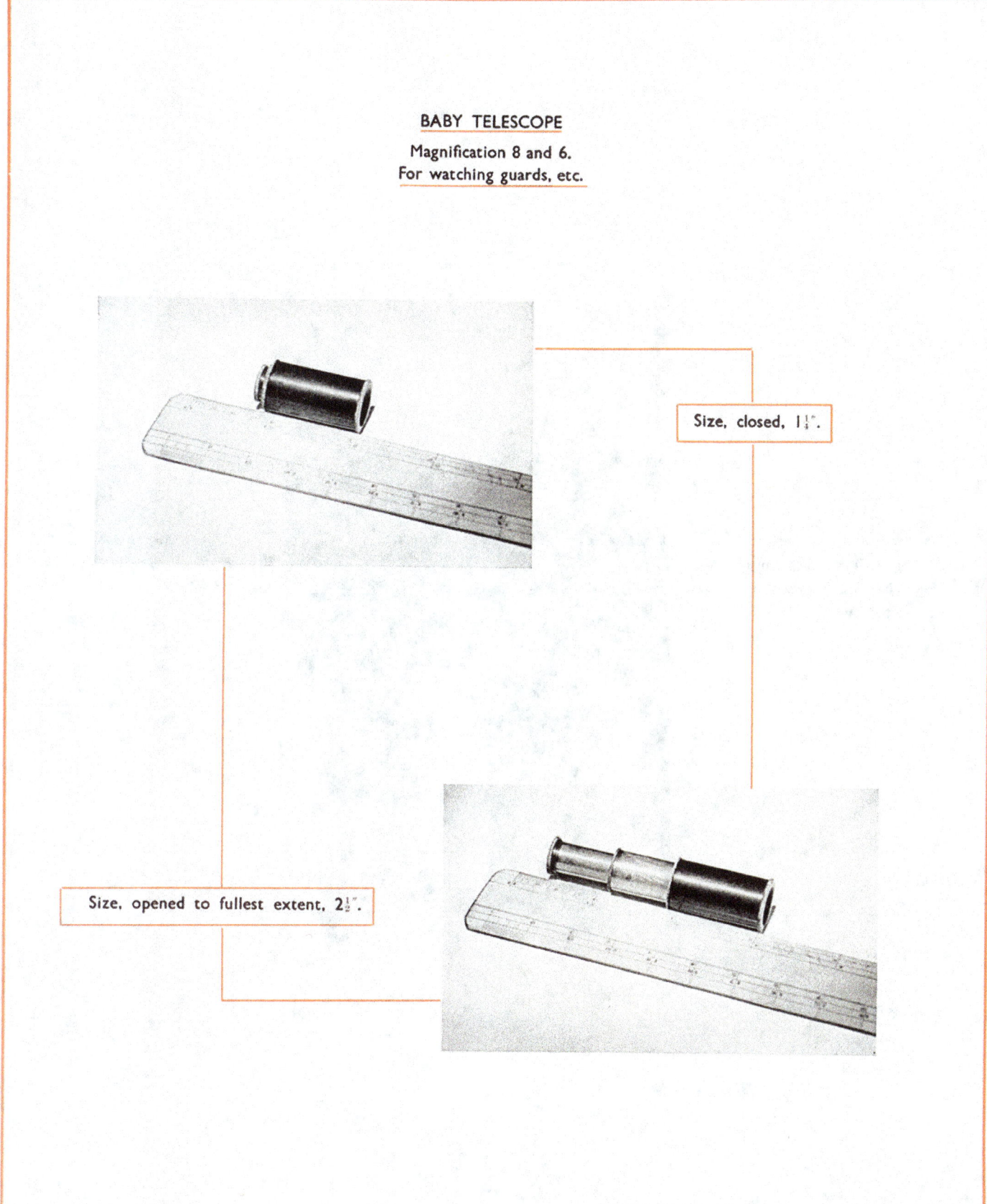

BABY TELESCOPE
Magnification 8 and 6.
For watching guards, etc.

Size, closed, $1\frac{1}{4}''$.

Size, opened to fullest extent, $2\frac{1}{2}''$.

POST CAPTURE

SPECIAL MESS DRESS.

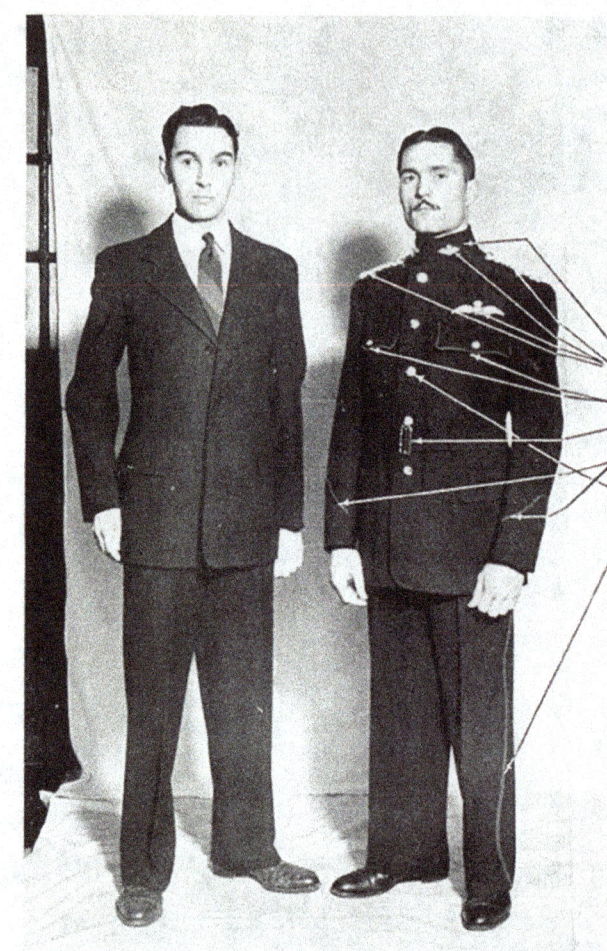

In one minute a perfect fitting walking out suit can be made — which is also waterproofed.

- Take out Buckram
- Take off Badges
- Tear off Pockets & Flaps
- Take off Belt
- Tear off Sleeve Piping
- Buttons replaced by those from inside Trouser Tops
- Tear off Stripes

ESCAPE AIDS

TWO-WAY R.A.F. SUIT
Officers and O/R's.

Turned inside-out, a perfect **walking** out suit is **made**.

POST CAPTURE

AIDS

WERE PACKED AND DESPATCHED
TO THE VARIOUS P/W CAMPS

•

THE NAMES AND ADDRESSES OF ASSOCIATIONS, SOCIETIES AND NUMEROUS PRIVATE INDIVIDUALS, ALL OF WHICH WERE FICTITIOUS, WERE USED IN EVERY INSTANCE, SOME OF WHICH ARE GIVEN ON THE FOLLOWING PAGES

•

*ALL THESE WERE, IN FACT,
M.I.9*

POST CAPTURE

"Good" parcels and "Naughty" parcels ready for despatch.

ANOTHER NAME FOR M.I.9

PRISONERS' LEISURE HOURS FUND

"The treasures to be found in idle hours only those who seek may find."
—Runyan

President:
B. ATTENBOROUGH, Esq.

Vice-Presidents:
Sir THOMAS BERNEY, Bart.
L. C. UNDERHILL, Esq.

Committee:
Lady D. BROWNE.
The Hon. Mrs. E. FREEMAN.
P. O. NORTON, Esq.
J. B. WORLES, Esq.

**66 BOLT COURT,
FLEET STREET,
LONDON, E.C.4.**

Hon. Treasurer:
E. TOWNSEND, Esq., C.A.

Hon. Secretary:
Miss FREDA MAFFIN.

Telephone:
CENTRAL 3951

12th MAY, 1941

Dear Sir,

    Through the kindness of one of our contributors, we are enabled to send to you a selection of Musical Instruments - and Gramophone Records, and we are having despatched direct from the manufacturers in the course of a few days some records.

    We intend despatching different selections for each prisoner of war - to whom we send these, and it is hoped in order that all may enjoy the variety, you will offer to interchange with each other.

    Further supplies will be sent you at regular intervals, and if there is any particular record you desire sent, perhaps you will look through the Catalogues we are sending letting us know the make and number, and we will do our best to despatch to you in due course.

    Trusting you are enjoying good health, and looking on the bright side of things.

    Yours faithfully,

    Secretary.

A Voluntary Fund formed for the purpose of sending Comforts, Games, Books, etc. to British Prisoners of War.

ANOTHER NAME FOR M.I.9

LICENSED VICTUALLERS SPORTS ASSOCIATION

(WHOLESALE ONLY)

Telephone:
CENTRAL 6952

10, St. BRIDE STREET,
LONDON, E.C.4.

Secretary:
J. H. SHERWELL

Suppliers of Games and Bar Requisites to Hotels, Restaurants, Sports Clubs and other Licensed Premises.

ACKNOWLEDGEMENTS

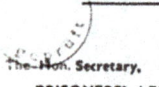
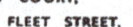

In order to test the receipt of parcels, a special card was inserted in each parcel, with a request in the accompanying letter that the German Camp Commandant would be good enough to allow this card to be returned — thus saving a weekly P/W letter.

Nearly all were returned.

A

ACKNOWLEDGEMENTS

Acknowledgements of "Dove" parcels which were always sent before "naughty" parcels.

ACKNOWLEDGEMENTS

Note
"Not Known"
— and request
on back of card.

D

ACKNOWLEDGEMENTS

Note acknowledgements of clothes from P.L.H.F., thus giving proof that the Red Cross was not the only Society "accepted" by the Germans.

ACKNOWLEDGEMENTS

Note different post marks of Camps.

56

ACKNOWLEDGEMENTS

Acknowledgements of "Naughty" records, (one "Naughty" record with every four "Doves"). (Page 43.)

ACKNOWLEDGEMENTS

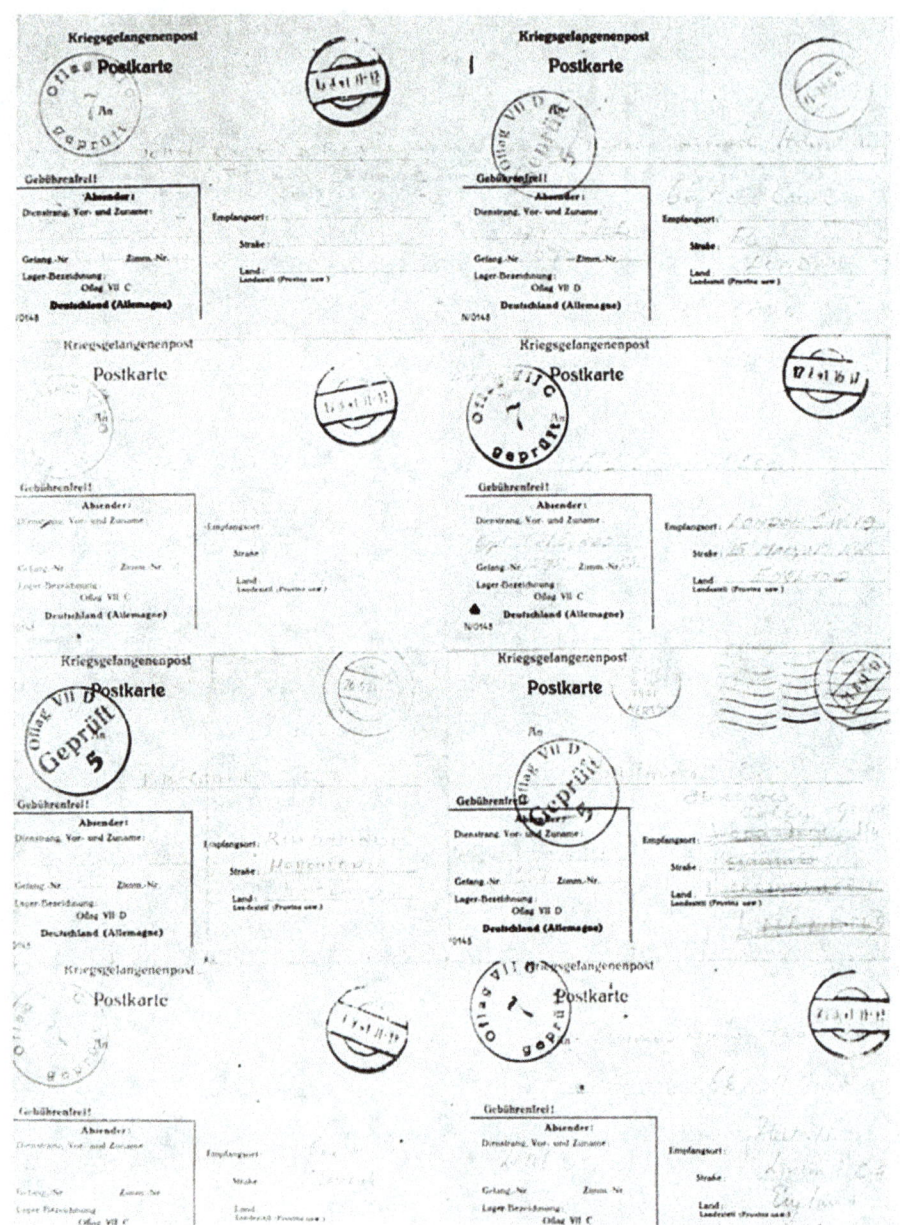

Acknowledgements to "Private individuals"— for parcels.
(All names and addresses were "manufactured" by M.I.9.)

ACKNOWLEDGEMENTS

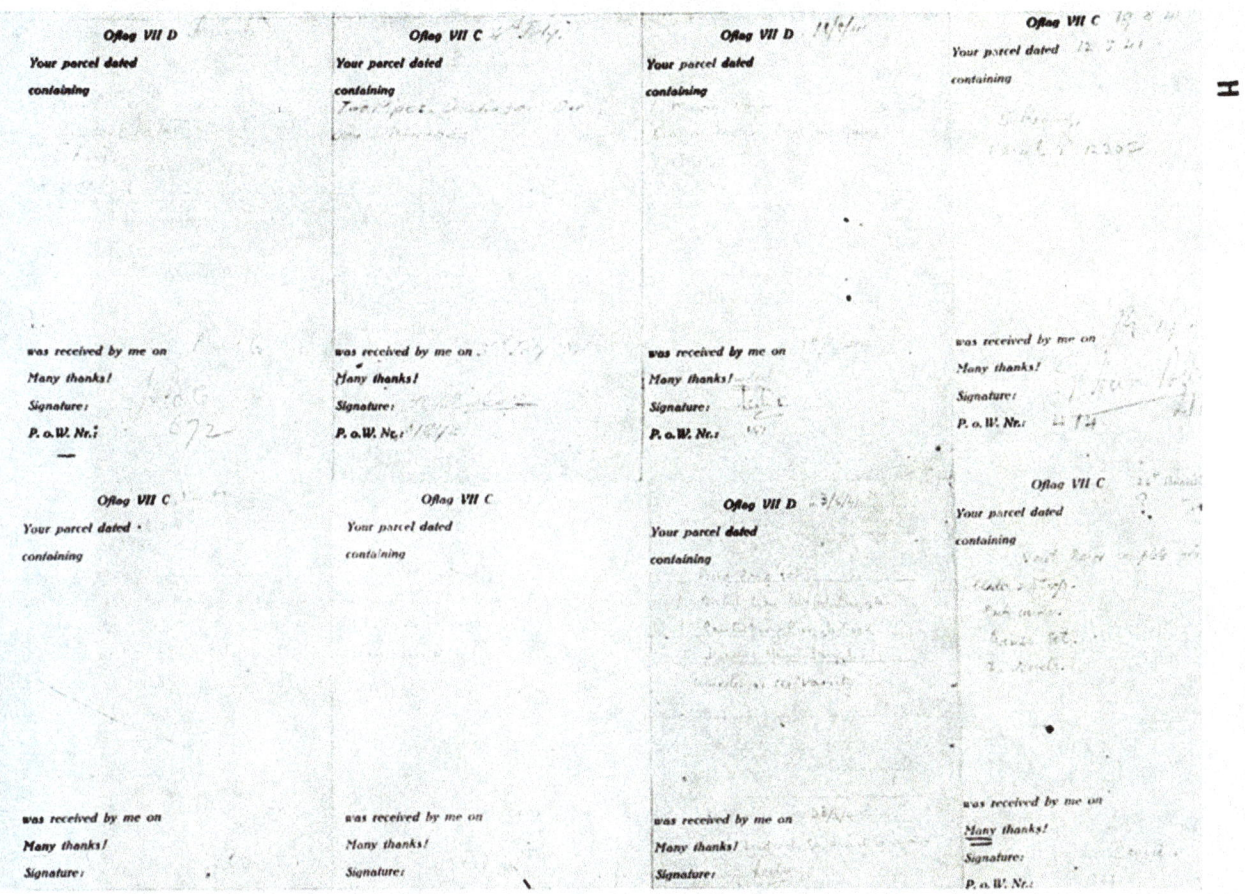

Acknowledgements of "Naughty" games, etc. (Note remarks!)

ACKNOWLEDGEMENTS

Acknowledgement of "Naughty" books.

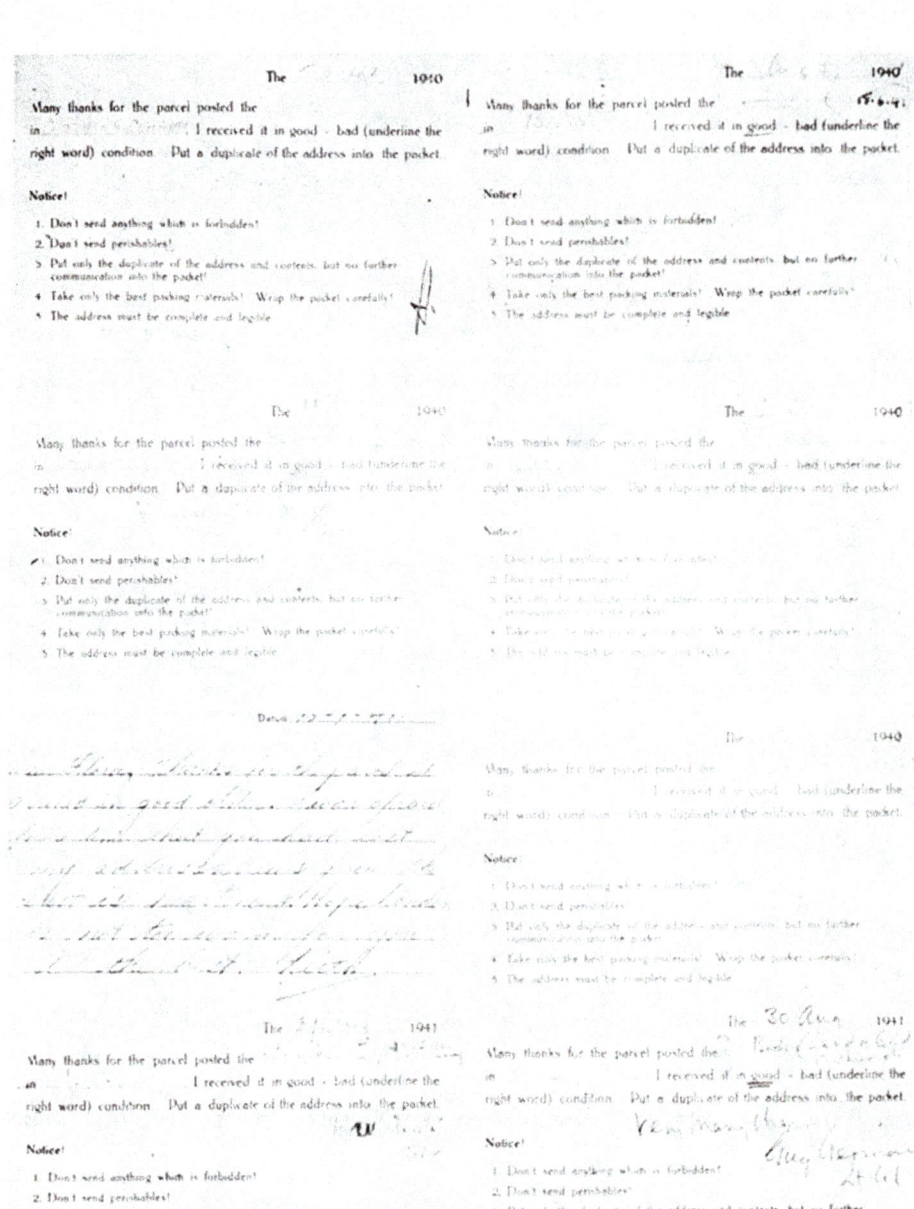

ACKNOWLEDGEMENTS

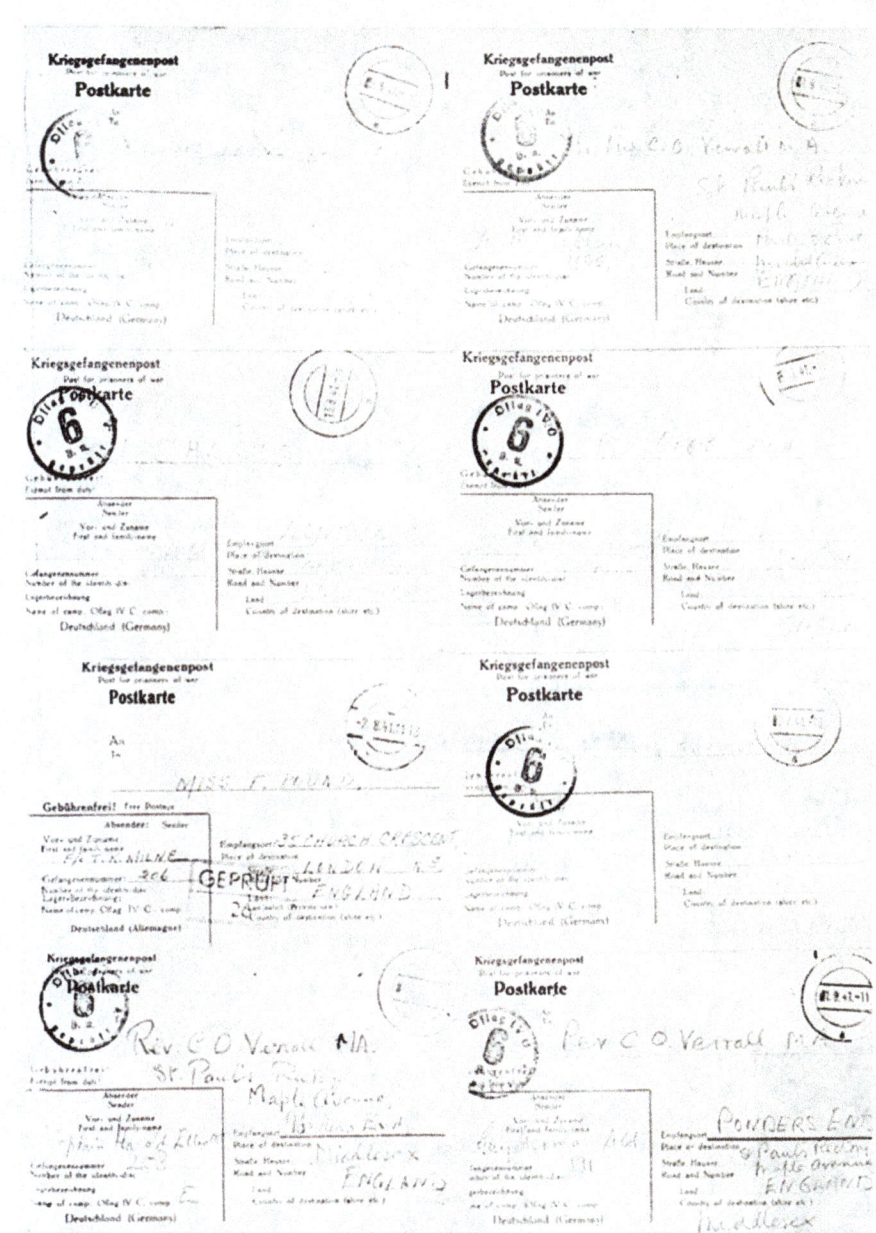

Acknowledgements of Games and *Books*.

Note:
The Rev. C. O. Verrall, M.A. (Cover all) had his vicarage bombed by the enemy! He accordingly sent all the remaining books from his very old library!

ACKNOWLEDGEMENTS

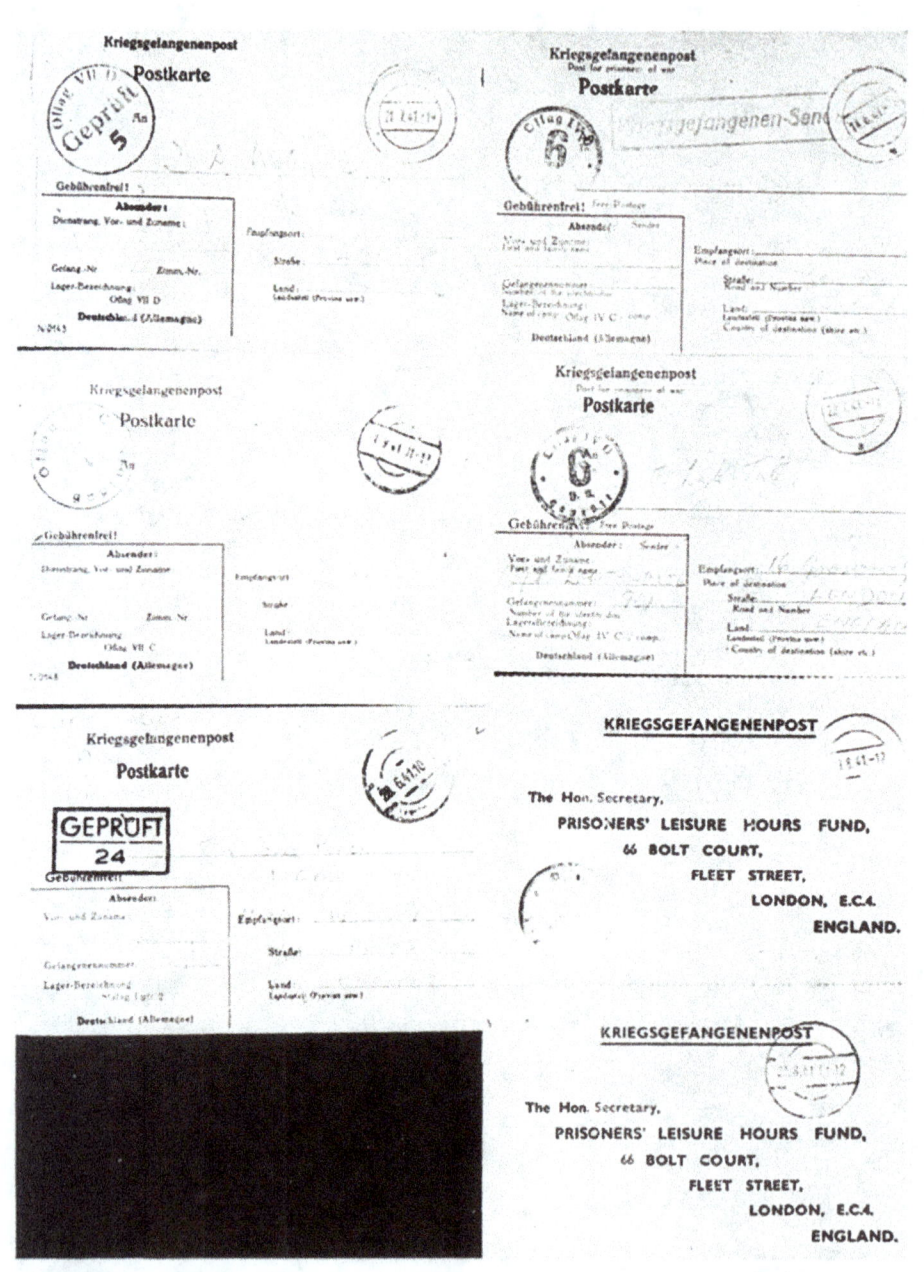

M.I.9
" Phoney addresses. "
Acknowledgements.

ACKNOWLEDGEMENTS

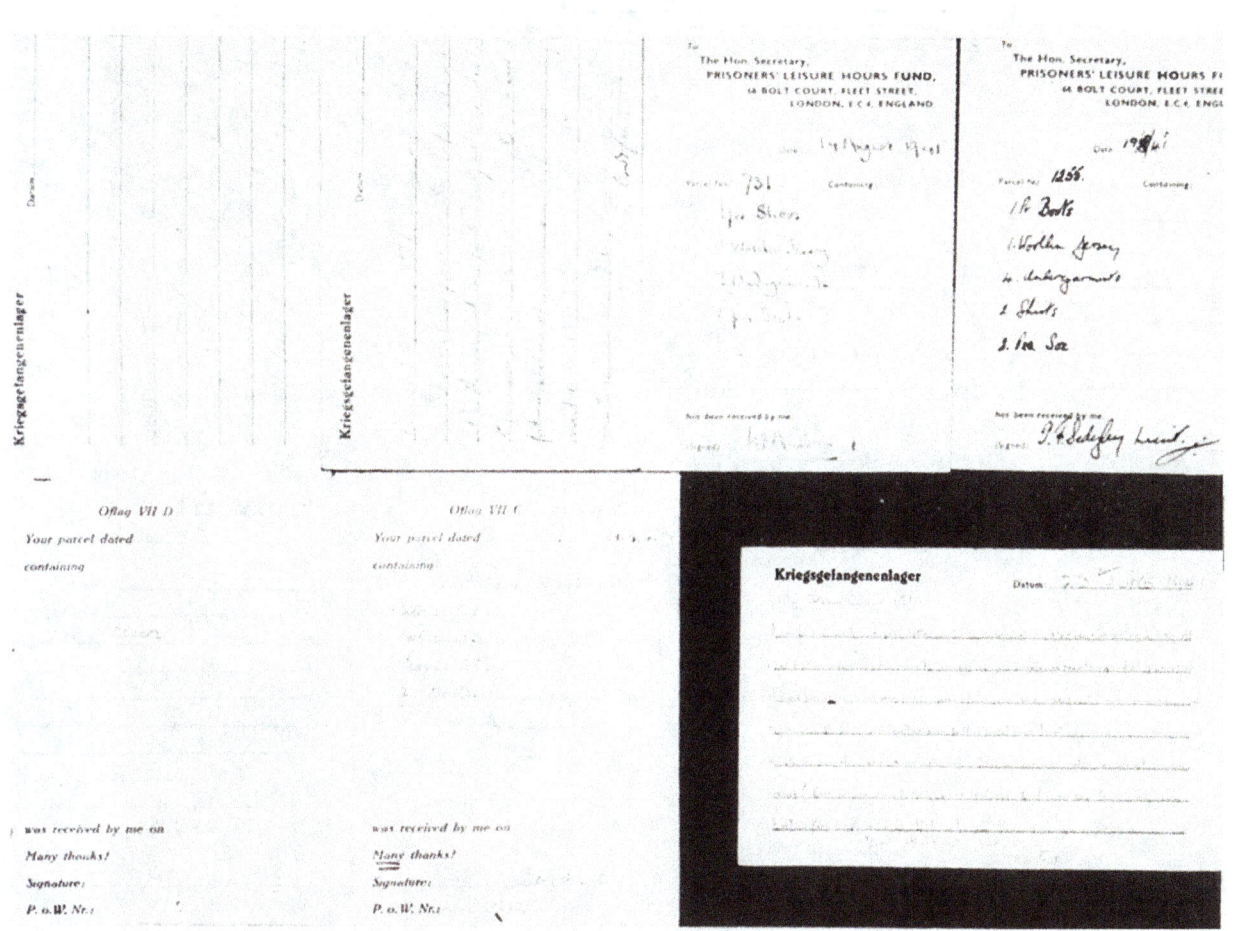

Acknowledgements in *clear* of "Naughty" Goods.

ACKNOWLEDGEMENTS

COPY POSTCARD FROM CAPT. G.F.K. DALY, SULMONA ITALY to THE HON. SECRETARY, PRISONERS' LEISURE HOURS FUND, 66, BOLT COURT, FLEET STREET, LONDON, E.C.4.

21/8/41.

Dear Secretary,

The words quoted at the head of your notepaper by G. Runyan are apt. Space is certainly rather limited. No grounds for useless discouragement, in fact! One is here a prisoner & so must meet the inevitable boredom with one's head high. The spiritual and material necessities supplied by you with such generosity make camp life very much easier to cope with. Already the fourth clothing parcel of the May lot has arrived. Some six weeks slower than usual, all the same they have got here. Quite safely, too, and complete.

I want to thank you, but find difficulty in expressing what it means to us to know we have such kind friends thinking of us at home. Major Pritchard, Lieut. Deane-Drummond and 2nd. Lieut. Paterson join me in my most sincere thanks for your very real kindness.

Yours Ever,

G.F.K. Daly.

Capt.

One of the many acknowledgements to M.I.9 indicating the activities were well known — and appreciated.

MESSAGE

PRIME MINISTER'S
PERSONAL MINUTE   10, Downing Street,
                   Whitehall.
SERIAL N: M

Message from the Prime Minister
to the Prisoners of War.

In this great struggle in which we are engaged, my thoughts are often with you who have had the misfortune to fall into the hands of the Nazi.

Your lot is a hard one, but it will help you to keep your courage up to know that all is well at home. Never has the country been so completely united in its determination to exterminate Nazidom and re-establish freedom in the world. Our strength grows daily, and assistance flows from America in ever-increasing volume. In high-hearted confidence we press forward steadily along the road to certain victory.

Keep yourselves fit in mind and body, so that you may the better serve our land, and, when peace comes, play your part in establishing a happier, safer homeland.

God bless you all.

Winston S. Churchill

August 5, 1941.

One of the many "invisible" messages sent on cotton handkerchief.
(See Page 42.)

ESCAPE AIDS

SPECIAL WIRELESS RECEIVERS

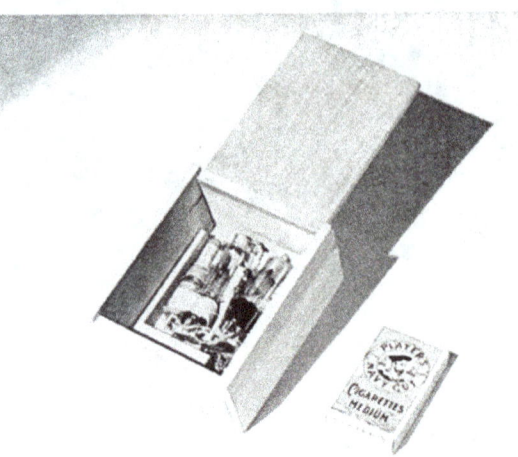

Cigar Box Type.   Range 400 miles.

6" x 6" x 1½".   Range 700 miles.

WIRELESS TRANSMITTERS

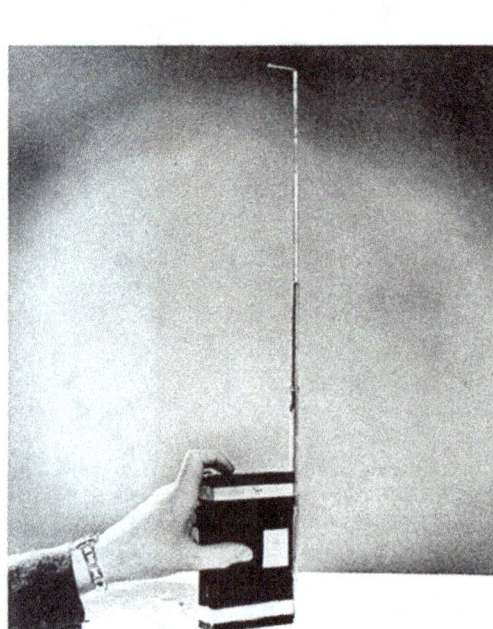

Note Telescopic 2' 6" Mast.

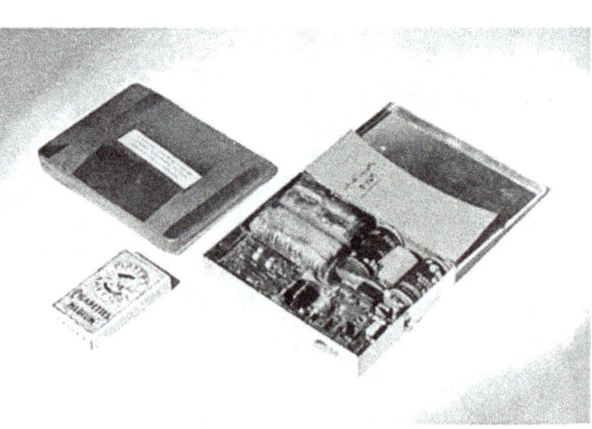

100 Players Cigarette Tin Type.
Range 100 miles.

ESCAPE AIDS

SPECIAL WIRELESS RECEIVERS.

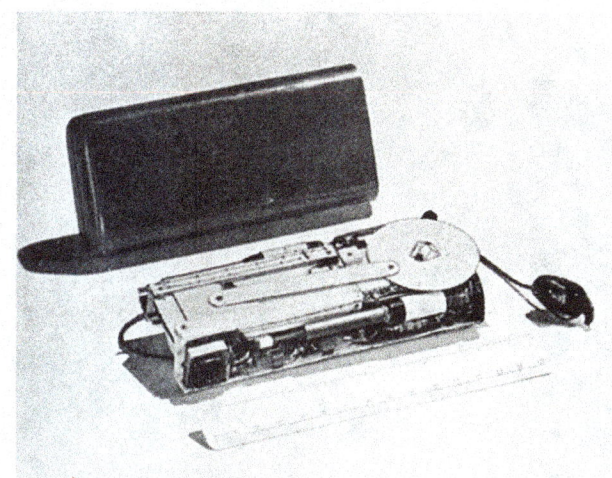

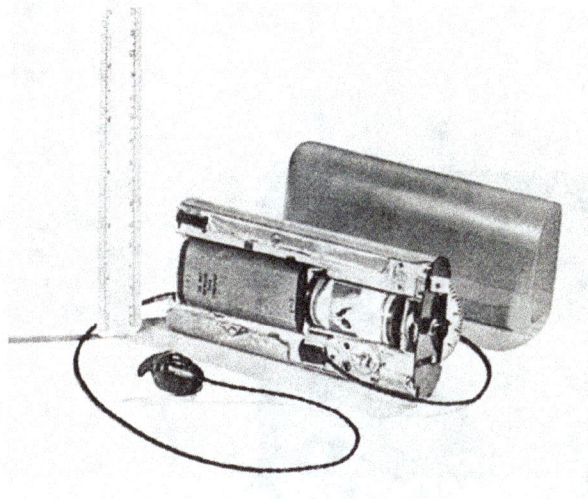

3 Cigar Case Size. Range 250 miles.

ESCAPE AIDS

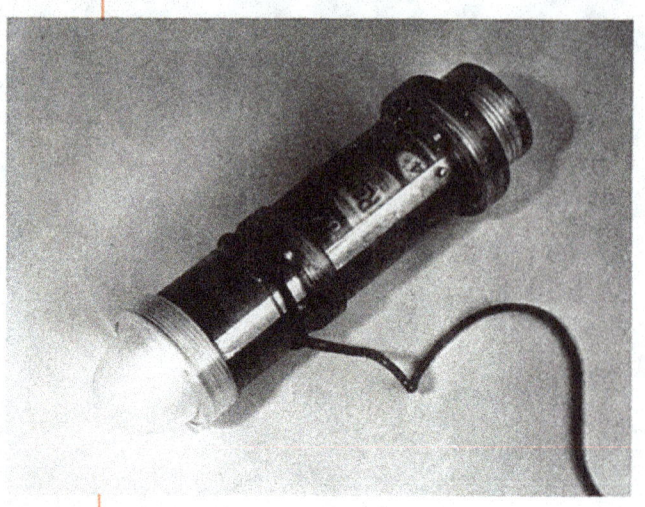

This Torch is constructed of acetate. It weighs 6 ozs. It is watertight.

It is bottom heavy. The user can either Morse if capable of doing so, or if overcome through fatigue or enemy action can, by turning bottom, keep torch alight for twenty-four hours. It is fitted with lanyard, and floats in water as shown in the lower illustration.

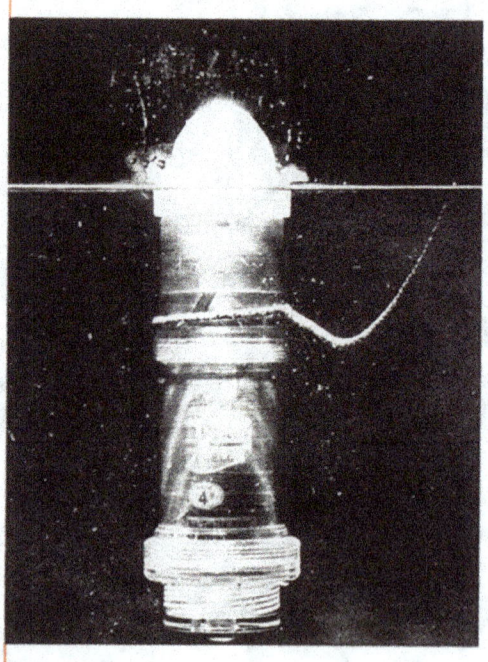

AIDS FOR OTHER SERVICE SECTIONS

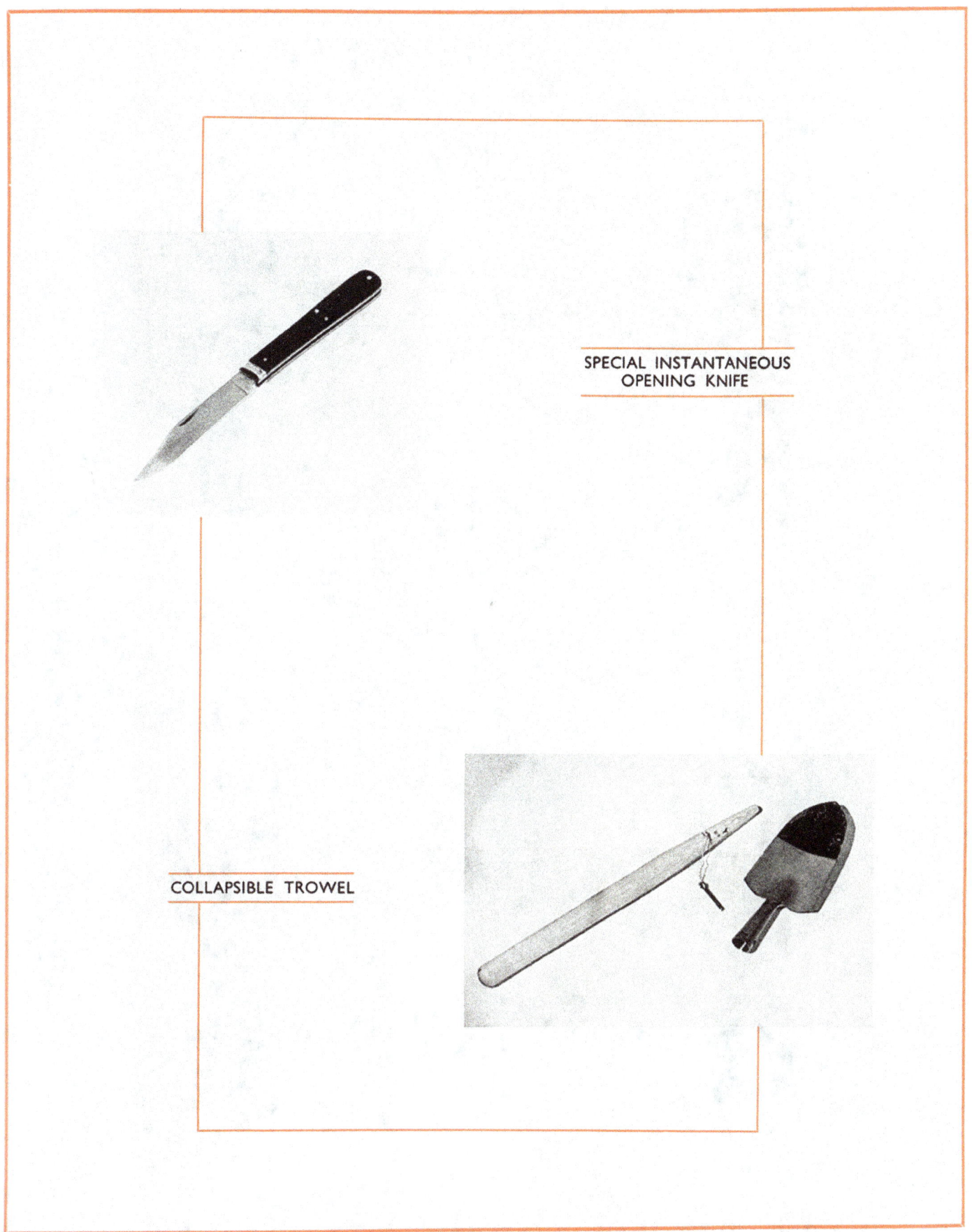

SPECIAL INSTANTANEOUS OPENING KNIFE

COLLAPSIBLE TROWEL

FOREIGN COSTUMES

GERMAN AIR FORCE OFFICER.

OTHER RANKS.

FOREIGN COSTUMES

GERMAN INFANTRY.

"TWO WAY" COAT
"Newmarket"

GERMAN STEEL HELMET
Exact in every detail.

FOREIGN COSTUMES

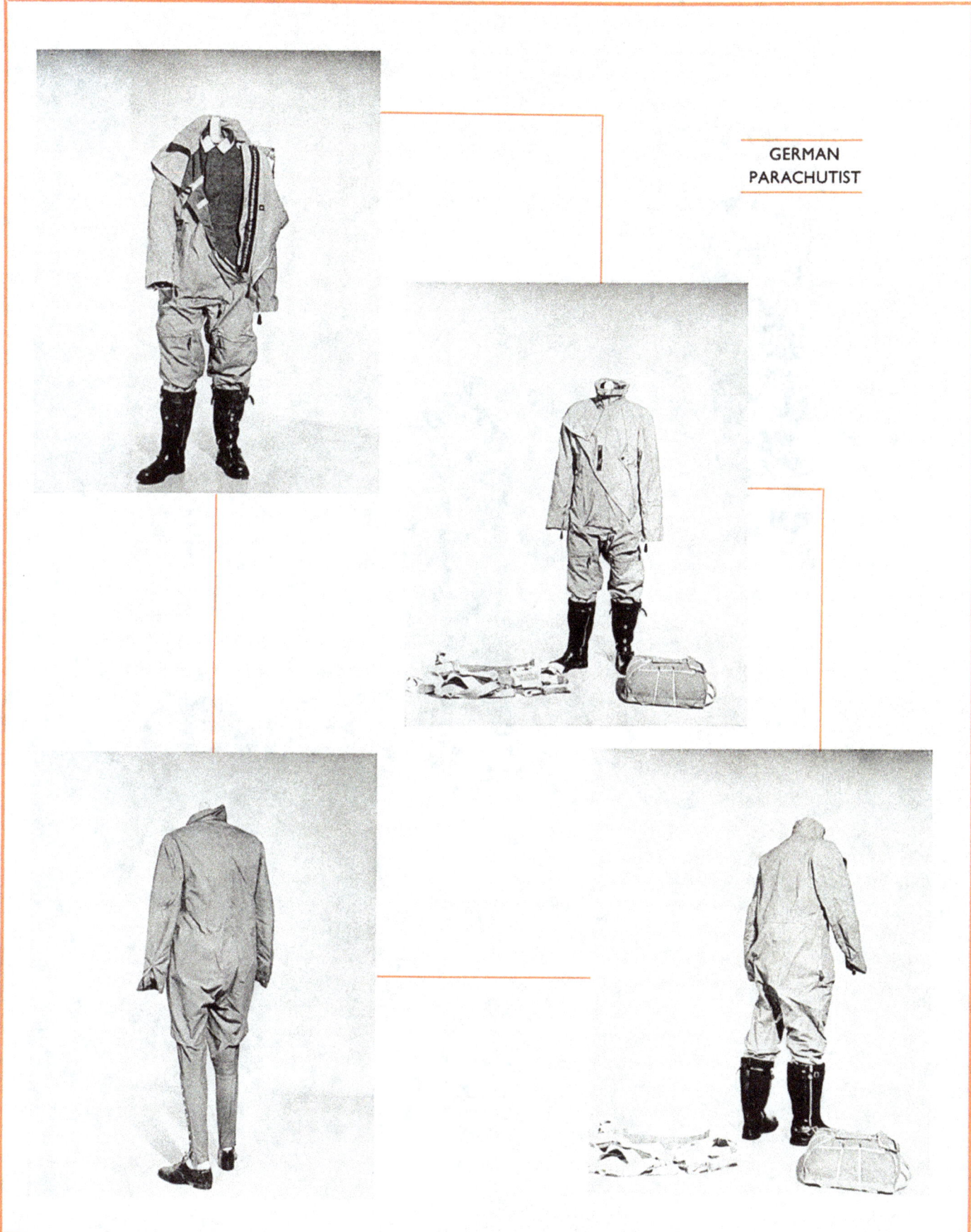

GERMAN PARACHUTIST

FOREIGN COSTUMES

NORWEGIAN

FOREIGN COSTUMES

SPANISH
TUNIC

Part of the functions of M.I.9 was to produce Foreign costumes — of all combatants.
These were produced in exact detail.
The copies could not be distinguished from the originals.

SPANISH
GREATCOAT

MINIATURE CAMERAS

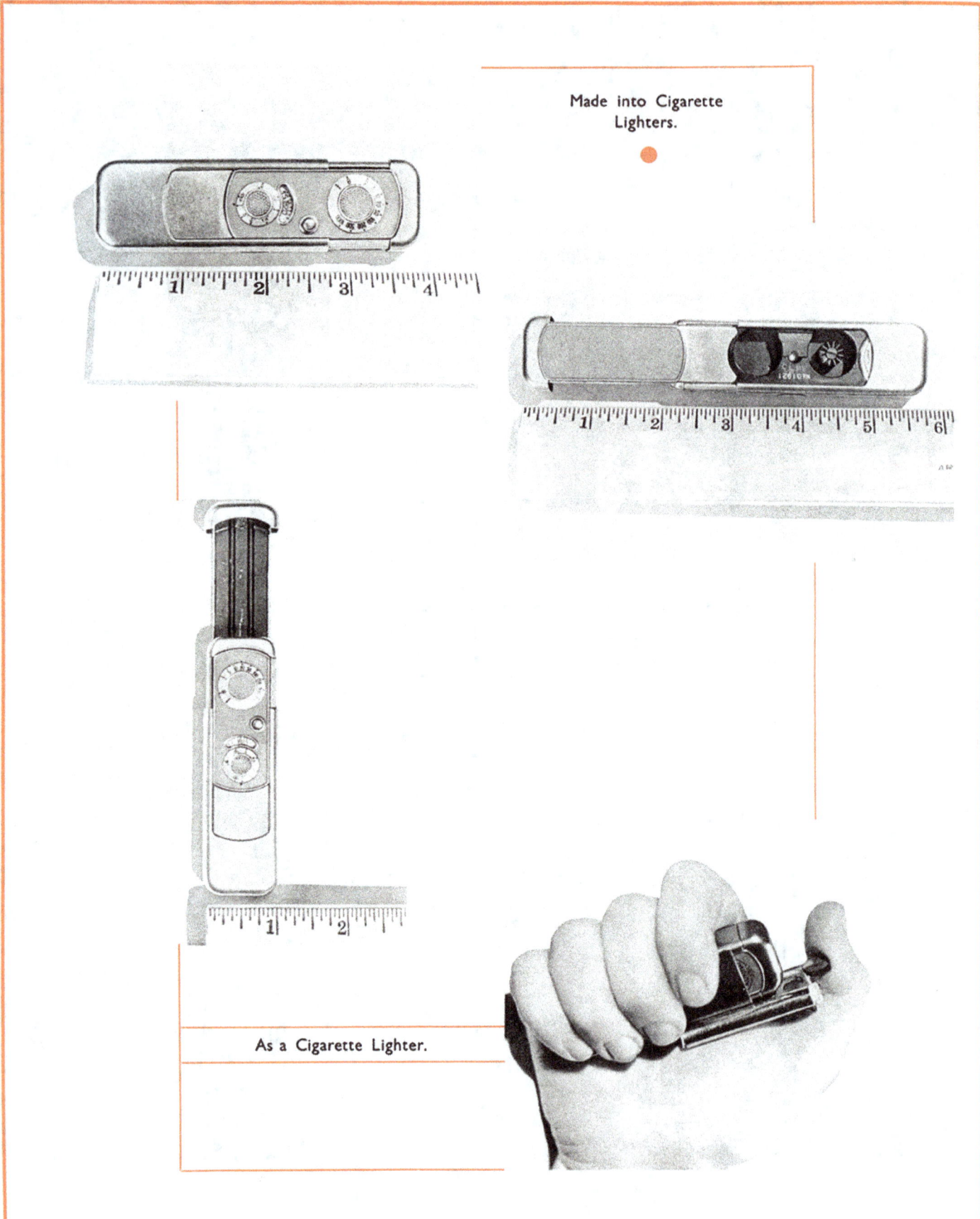

Made into Cigarette Lighters.

As a Cigarette Lighter.

www.ingramcontent.com/pod-product-compliance
Lightning Source LLC
Chambersburg PA
CBHW080027130526
44591CB00037B/2703